Investigating Life in the Universe

This textbook gives a lively introduction to the search for extraterrestrial life. It is a guidebook to understanding the possibility of life elsewhere, pointing out landmarks and providing background information to facilitate further exploration of those areas of most interest to the reader.

We are a planet of winners – winners of a cosmic lottery that has been in play since the universe began approximately 13.7 billion years ago. Our winnings include sentience and an underlying unease that has driven us to contemplate our place in the universe and the possibility of finding kindred spirits in the cosmos spreading out before us. To understand our origins and the possibility of life beyond Earth, we must look back and retrace the steps that have brought us to this point in space and time. In doing so, we will find the investigation of life to be a unifying theme in nature, requiring us to touch on all branches of the tree of knowledge. Using the Drake Equation as a theme, we begin with an overview of the topic and then go into the story of how we have acquired, or plan to acquire, the knowledge to solve it. As we make our journey, we will encounter some very interesting people – some you will likely know, while others may be new to you. Keep an open mind and allow this text to be your guide.

Written in an engaging style, this textbook provides a foundational understanding of the rapidly advancing fields associated with the search for life in the universe. Each chapter includes illustrative figures and review questions for self-study. It will appeal to professionals, researchers, instructors, and undergraduate students, as well as anyone with an interest in astrophysics or astrobiology.

Investigating Life in the Universe

Astrobiology and the Search for Extraterrestrial Life

Christopher K. Walker

CRC Press is an imprint of the
Taylor & Francis Group, an **informa** business

Cover Image Credits:
JWST Deep Field Image: NASA, ESA, CSA, and STScI
JWST Image: NASA
Perseverance Rover Image: NASA/JPL
Trappist-1 System: NASA/JPL-Caltech
Trappist-1e Model Spectrum: Dr. Carrie Anderson, GSFC
VLA Image: NRAO/Jeff Hellerman
Giant Magellan Telescope: GMTO Corporation

First edition published 2024
by CRC Press
2385 NW Executive Center Dr. Suite 320, Boca Raton, FL 33431

and by CRC Press
4 Park Square, Milton Park, Abingdon, Oxon, OX14 4RN

CRC Press is an imprint of Taylor & Francis Group, LLC

ISBN: 9781138628717 (hardback)
ISBN: 9781032472522 (paperback)
ISBN: 9781315210643 (eBook)

DOI: 10.1201/9781315210643

Typeset in Minion
by Deanta Global Publishing Services, Chennai, India

Dedication

To my students and colleagues, who have made this academic journey possible.

Contents

Preface

There are more things in heaven and earth, Horatio,

Than are dreamt of in your philosophy.

– *HAMLET* (1.5.167–8), HAMLET TO HORATIO

WE ARE A PLANET of winners – winners of a cosmic lottery that has been in play since the universe began ~13.7 billion years ago. Our winnings include sentience and an underlying unease that has driven us down the tortuous path of evolution to the point of contemplating our place in the universe and the possibility of finding kindred spirits in the cosmos spreading out before us.

Where did we come from? Where are we going? What does it mean to be alive? Are we alone? These are questions that have occurred to each of us and are the corner stones of our investigation of life in the universe. We are the products of astrophysical, chemical, biological, cultural, and technological evolution. For the moment at least, the only instances of life we know of with certainty are associated with the Earth. However, scientific inquiry has shown us the same physical laws at work here also apply to the most distant parts of the observable universe. Therefore, there is no physical reason to assume the occurrence of life is unique to the Earth. To understand our origins and the possibility of life beyond Earth, we must look back and retrace the steps that have brought us to this point in space and time. In doing so, we will find the investigation of life to be a unifying theme in nature, requiring us to touch on all branches of the tree of knowledge. By seeing the interconnection between what at first appears to be disparate fields of inquiry, we will begin to see the weave of the universe and, perhaps, gain a better perspective of our place in it.

This book is meant to serve as an introduction to the search for extraterrestrial life. This is a rapidly advancing field. Indeed, when the author began teaching a course on extraterrestrial life ~25 years ago, there were not yet any definitive detections of extrasolar planets. As of today, there are over 5,000 confirmed planets, some of which, from the data available, may be Earth-like. In the past quarter of a century, there have also been advances in the associated fields of astronomy, planetary science, evolutionary biology, archeology, and neural science. The author has worked to utilize much of his existing course material in preparing this book, but found in the light of new discoveries, that some of it was dated, incomplete, or just wrong. The author has done his best to make the material as up-to-date as possible at the time of this writing. At best, this text is a guidebook to your journey to understand the possibility of life elsewhere. We begin with an overview of the topic in the context of the Drake Equation and then go into the story of how we have acquired or plan to acquire the knowledge to solve it. As we make this journey, we will encounter some very interesting people, some you will likely know, and others may be new to you. Behind each discovery, we will find the faces and names of those who pioneered it. If you find a particular individual or topic of interest, you are encouraged to make use of the provided references to dig deeper. As in the case of a guidebook, this text only points out some interesting landmarks along the way. It is up to the traveller to explore those areas of most interest to them. I wish you well on your journey.

Christopher K. Walker
Tucson, Arizona
July 23rd, 2023

Author

Professor Christopher K. Walker has over 30 years of experience designing, building, and using state-of-the-art receiver systems for terahertz astronomy. He is a Professor of Astronomy, Optical Sciences, Aerospace and Mechanical Engineering, and Applied Mathematics, and an Associate Professor of Electrical and Computer Engineering at the University of Arizona (UofA), USA. He has published numerous papers on star formation and terahertz (THz) technology. He has served as dissertation director for a number of PhD students and been a topical editor for *IEEE Transactions on Terahertz Science and Technology*. While working on his master's in Electrical and Computer Engineering at the Ohio State University, he was chief engineer of the "Big Ear" radio telescope, which conducted one of the first all-sky searches for extraterrestrial intelligence (SETI). Upon completing his MS, he worked at TRW Aerospace on microwave satellite systems and then at the Jet Propulsion Laboratory (JPL) on SETI and interplanetary radar. He then earned a PhD in Astronomy from the UofA. As a Millikan Fellow in Physics at Caltech, he worked on the development of low-noise, SIS waveguide receivers above 400 GHz and explored techniques for etching waveguide out of silicon. On joining the UofA faculty in 1991, he began the Steward Observatory Radio Astronomy Lab (SORAL), which has become a world leader in developing THz receiver systems for astronomy and other remote sensing applications. Instruments developed by Walker's team have served as primary facility instruments at the Heinrich Hertz Telescope on Mt. Graham, AZ, and the AST/RO telescope at the South Pole and have been used at the APEX telescope on the Chajnantor plateau in Chile's Atacama region. Walker led the effort to design and build the world's largest (64 pixels) submillimeter-wave heterodyne array receiver. He was Principal Investigator (PI) of the NASA long-duration balloon project, the Stratospheric THz Observatory (STO), which flew in Antarctica in 2012 and 2016. He is now PI of GUSTO, a balloon-borne mission under the auspices of the NASA Explorer Program and CatSat, a 6U CubeSat that demonstrates a new deployable antenna technology. Walker is co-founder of FreeFall Aerospace, which commercializes this technology. He was selected by the NASA Innovative Advanced Concepts (NIAC) program for both a Phase I and Phase II study. He authored *Terahertz Astronomy*, the first textbook in his field of study. He was named a Galileo Circle Fellow in 2015 by the College of Science of the University of Arizona. He has a loving wife and two children who have supported him through many observing runs, Antarctic deployments, and late nights.

CHAPTER 1

Drake Equation

An Overview

PROLOGUE

The possibility of life beyond the confines of Earth has been discussed since antiquity, sometimes at great peril. In 1600 Giordano Bruno was burned at the stake in Rome for, in part, espousing the belief that the universe was infinite and populated by many worlds. A decade later, Galileo Galilei constructed his first telescope and came close to suffering the same fate. It was not until the nineteenth and twentieth centuries that the technologies and scientific tools needed to rigorously explore the possibility of life beyond Earth came into existence. These discoveries led to the pivotal 1959 paper by Giuseppe Cocconi and Philip Morrison in which the authors discussed the possibility of communicating with extraterrestrial life using radio telescopes. A year later, Frank Drake led the first effort, called Project Ozma, to detect such signals using facilities at the National Radio Astronomy Observatory in Green Bank, West Virginia. As part of the preparation for a follow-up meeting on Project Ozma, he developed his now famous "Drake Equation" for estimating the number of extraterrestrial communicable civilizations, N_c. He did so by multiplying together key factors he believed were necessary for the occurrence of a communicable civilization. Initial estimates of N_c were based on limited knowledge of these underlying factors. Over the past half century, much progress has been made in refining these estimates and we are now able to more accurately estimate the possibility of life elsewhere in the Milky Way and beyond. In particular, rapid improvements in the technology and observing techniques associated with planet detection now enable us to make much more reliable estimates of the number of "Earth-like" planets in our galaxy. In the coming years, advances in the study of the origin of life and the search for life elsewhere within the solar system will reduce the error bars in calculations of N_c even further. The first definitive detection of an extrasolar planet orbiting a main sequence star was not until 1995. As of today, there are over 4,000 confirmed planets, some of which, from the data available, may be Earth-like. In this chapter we will take a walk through the Drake Equation and identify the factors that drive us to estimates of N_c. In the process, we will find the investigation of the possibility of life elsewhere in the universe to be a unifying theme in nature, requiring us to touch on all branches of the tree of knowledge.

DOI: 10.1201/9781315210643-1

1.1 INTRODUCTION

Our home within the local universe is the Milky Way galaxy, a rotating assemblage of ~400 billion stars, with a disk and a central nuclear bulge (see Figure 1.1a). The morphological appearance of the Milky Way is similar to that of a "Frisbee" with a golf ball stuck in the middle. Recent observations suggest the disk of the Milky Way is between 170,000 and 200,000 light years (ly) in diameter, with an average thickness of ~1,000 ly (Rix and Bovy 2013). The central bulge is ~20,000 ly in diameter. Our solar system is located in the galactic disk, about 27,000 ly from the galactic center. From our location in the disk, we see the Milky Way edge-on. On a clear, moonless night, it will appear to us as a diffuse band of light arching across the sky from horizon to horizon (see Figure 1.1b). The diffuse band is caused by the superposition of light from billions and billions of stars within the disk. From very dark locations, shadowy features can be observed along some lines of sight through the disk. These features are due to interstellar dust clouds that are sufficiently thick to diminish the light from the stars behind them.

Given this situation, it would at first appear estimating the number of advanced civilizations within the Milky Way to be a hopeless task, with the prospect of making such an estimate for a universe composed of ~200 billion galaxies being beyond imagination. *This is simply not the case.* If we assume the laws of nature operating here on Earth are the same elsewhere (which from all indications appears to be true), then we can use our understanding of these laws, together with observation, to estimate both the number and distance to our extraterrestrial neighbors in the Milky Way and beyond.

The Drake Equation is a concise, mathematical expression through which we can manifest our understanding of how we arrived at our current state of development on Earth for the purpose of estimating the number of communicable civilizations within a given

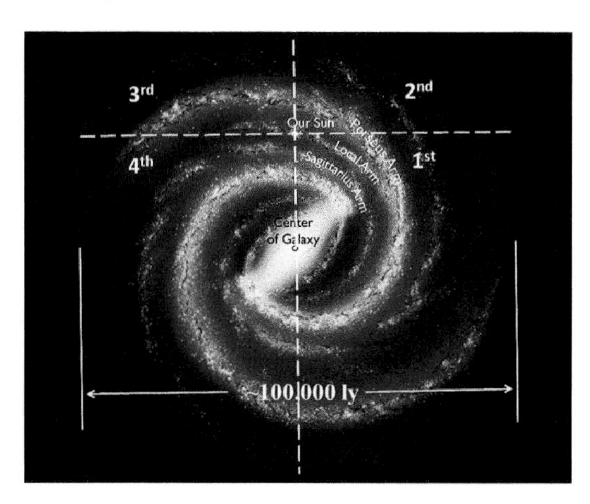

FIGURE 1.1A Face-on view of the Milky Way. Our galaxy is a Class SBc, barred-spiral galaxy approximately 100,000 ly in diameter, containing ~400 billion stars. The Sun is located ~27,000 ly from the center. The central bulge or nucleus of the Galaxy containing the bar is ~10,000 ly in diameter. The galaxy is divided into four quadrants, with our solar system at the center. Underlying Milky Way image credit: NASA. https://www.nasa.gov/feature/jpl/astronomers-find-a-break-in-one-of-the-milky-way-s-spiral-arms

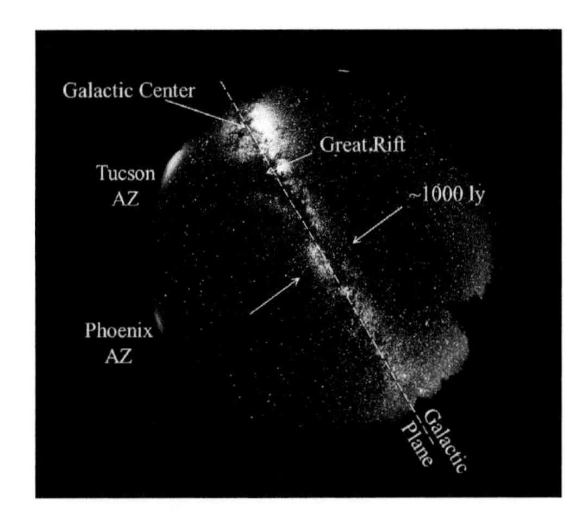

FIGURE 1.1B Edge-on view of the Milky Way as seen from the summit of Mt. Graham, Arizona. The Galactic Plane appears ~30° in width and is the home of numerous, overlapping interstellar dust clouds (e.g., the Great Rift), clear evidence of an extensive interstellar medium (ISM). The Galactic Center is along a line of sight toward the brightest region of the Milky Way. Underlying Milky Way photo: Copyright 1997, Steward Observatory, The University of Arizona. Used with permission.

volume of space, e.g., the Milky Way. Here, "communicable" means having the ability and desire to search for and communicate with extraterrestrial species. In its most basic form, the Drake Equation can be expressed as

$$N_c = R_* f_g f_p n_e f_l f_i f_c L_c \tag{1.1}$$

where

R_* = the rate of star formation ($M_\odot$ per year)
f_g = fraction of stars that are "good" stars
f_p = fraction of stars that form planets
n_e = number of planets per stellar system that are "Earth-like"
f_l = fraction of planets where life originates
f_i = fraction of planets with life that evolves an intelligent species
f_c = fraction of intelligent species that have the desire and ability to engage in interstellar communications
L_c = lifetime of the interstellar communicative phase of a species (in years).

In order to assign values to these parameters requires a basic understanding of the processes leading to our existence, as well as an understanding of the technologies necessary to extend our reach to the stars. These processes can be cast in terms of physical, biological, cultural, and technical evolution (see Figure 1.2). Rewriting Eq. (1.1) with this in mind yields,

$$N_c = \text{physical evolution} \times \text{biological evolution} \times \text{cultural evolution} \times \text{technical evolution} \times L_c \tag{1.2}$$

where

physical evolution $= R_* f_g f_p n_e$

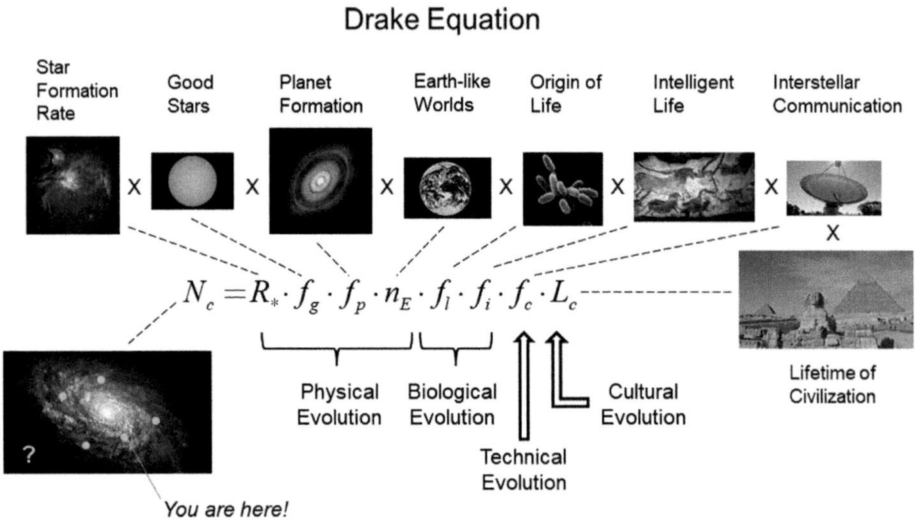

FIGURE 1.2 Drake Equation. The equation is a concise, mathematical representation that can be used to estimate the number of extrasolar species currently capable of interstellar communications within a designated volume of space, typically a galaxy. The equation's parameters can be categorized to represent various aspects of physical, biological, technical, and cultural evolution that will lead to the emergence of a species capable of interstellar communications. The greatest uncertainties in the equation are associated with the origin and evolution of life, for which the only example available to human investigators is what occurred on Earth. Nonetheless, when representative values based on knowledge of astronomy, geology, biology, and human evolution are substituted into the equation, the result suggests that whether or not other communicative species currently exist in the Milky Way is a sensitive function of the average lifetime of technically advanced civilizations. Image credits: a) Milky Way [The Hubble Heritage Team/AURA/STScI/NASA]; b) Orion Nebula [NASA]; c) Sun [NASA]; d) HL Tau [NRAO]; e) Earth [NASA]; f) Halobacterium [NASA]; g) Cave Painting [Shutterstock]; h) Antenna [Shutterstock].

biological evolution $= f_l f_i$
cultural and technical evolution $= f_c$.

As can be seen from Eq. (1.2), to address the questions related to the possibility of life elsewhere in the universe and our prospects of communicating with it requires insight into much of the human condition. In the next section, we will perform a walk through of the parameters in the Drake Equation. In later chapters, we will take a deeper dive into the science and technologies required to estimate each parameter.

1.2 DRAKE EQUATION PARAMETERS

1.2.1 Star Formation Rate: R_*

We are children of the Earth. In turn, the Earth and all the planets are byproducts of the processes leading to the formation of the Sun (see Figure 1.3). The Sun also provides the warmth to nurture and sustain life on Earth. If we assume life elsewhere developed in a similar fashion, then it makes sense that the star formation rate, R_*, is the first parameter in the Drake Equation. We can arrive at an initial estimate of R_* simply by taking the total

FIGURE 1.3 Orion Nebulae. Visible to the naked eye, the Orion Nebulae is a massive cloud of gas and dust located ~1,344 ly from the Earth. It is an active region of star formation and home to over 3,000 young stars. The Sun was formed from the gravitational collapse of a gas clump within such a cloud. Image Credit: https://hubblesite.org/contents/media/images/2006/01/1826-Image.html

number of stars in the Milky Way, N^*, and dividing it by the Age of the Milky Way, τ_{MW}. As stated above, the value of N^* is currently estimated to be ~ 400 billion. From the ages of stars, the Milky Way is believed to have been formed ~500 million years after the Big Bang. This puts the value of τ_{MW} at approximately 13.2 billion years. Dividing these two quantities yields an average value for R_* of ~30 stars per year. The star formation rate within the Milky Way was greater in the past than now. The current best estimate for R^* is four stars per year.

1.2.2 Fraction of Stars that are "Good": f_g

Here we define a "good" star as one that can serve as a stable source of light and heat over the many thousands of eons required to nurture the origin and evolution of intelligent life on one of its orbiting worlds. While the Earth formed ~4.7 billion years ago, intelligent life on Earth (namely, us!) with the desire and ability to communicate between the stars is a very recent occurrence, within the last ~50 years. This means it took essentially the entire time the Earth has existed to get us to this evolutionary threshold. There were many fits and starts along the way, where life on Earth was on the verge of total annihilation. Therefore, our best chance of finding life elsewhere is likely to be on planets orbiting stars that are expected to live at least as long as our Sun. The life span of a star is determined by its mass. Our Sun is

FIGURE 1.4 The Sun. The epitome of a "good" star, our Sun is a main sequence, G2 dwarf star that has provided a stable source of energy for the evolution of life on Earth for over 4.7 billion years. Image Credit: Shutterstock.

defined to be one solar mass, designated as $1M_{\odot}$. Across the Milky Way, stars are found with masses ranging from ~$0.01M_{\odot}$ to $150M_{\odot}$. Based on their mass and temperature, stars are assigned spectral classes, designated by letters: O, B, A, F, G, K, and M, with O stars being the most massive and M stars being the least. These classes are further divided by numbers 1–9. Our Sun (see Figure 1.4), for example, is classified as a main sequence, G2 dwarf star. A main sequence star is one in which nuclear fusion is transmuting hydrogen into helium in its core. It is the waste energy from this nuclear reaction that powers the Sun. Somewhat ironically, we will learn that due to the temperature sensitivity of fusion reactions, the lifetime of a main sequence star is inversely proportional to its mass; i.e., higher mass stars burn both brighter and faster than lower mass stars, making stars more than a few times more massive than our Sun unsuitable for sustaining life over the long timescales required to achieve intelligence. The good news is that conditions within the interstellar medium are such that many more low-mass stars, like our Sun, are formed than high-mass stars. However, the lowest-mass stars, while extremely long lived, are unstable, causing their energy output to vary significantly over time. This makes them ill-suited to provide the constant nurturing environment required for the evolution of intelligent life. We shall also learn that stars often form in clusters, leading to the occurrence of multiple star systems. Due to the complexities of the gravitational fields within such systems, the orbits of planets associated with them may be highly elliptical or convoluted; at one point bringing them too close to a star and later too far away to provide the relatively narrow temperature range (~100°C) required for the evolution of life. Taking these factors into account yields a value for f_g of ~0.12.

1.2.3 Fraction of Stars that have Planets: f_p

The idea that stars are distant suns that could harbor planets with life has been discussed since at least the time of Aristarchus of Samos (310–230 BC), an ancient Greek astronomer and mathematician. Only a few scraps of writings from this time period survived the destruction of the Great Library at Alexandria (around 275 AD) and the Dark Ages that followed. Aristarchus' work on the heliocentric model of the solar system was rediscovered during the Renaissance and used by Copernicus to develop a mathematical model for predicting the movements of the planets around the Sun. Copernicus waited to publish his work until just before his death in 1543 AD to avoid possible persecution from the Catholic Church, which considered the theory blasphemous. Institutional animosity against the heliocentric theory persisted for some time. In 1600 AD the Italian Dominican friar Giordano Bruno was burned at the stake by the Roman Inquisition (a fate dodged by Copernicus and, later, Galileo) at least in part, for openly proclaiming his belief that the stars are far-away suns that could potentially have inhabited planets orbiting them. This belief has come to be known as cosmic pluralism. He also believed the universe was infinite and had no center. He was close to getting this right as well. In the coming centuries, cosmic pluralism became more widely accepted, but it was not until 1992 AD that the presence of extrasolar planets (or exoplanets, for short) was observationally confirmed. There are currently several approaches used by astronomers to detect extrasolar planets (see Chapter 8). The most successful technique involves looking for periodic dimming of a star's brightness due to one or more of its planets periodically blocking a tiny fraction of its light as the planet passes between the star and the observer. This approach is referred to as the transit technique and has been used with great effect by the Kepler Space Observatory, which confirmed the presence of 2,662 planets. Based on these results, as well as those from other observatories, the percentage of stars harboring planets is currently estimated to be 70%, yielding a value for f_p of 0.7 (Figure 1.5).

1.2.4 Number of Earth-like Planets: n_e

Earth is the only planet we know of that has provided the conditions necessary for the origin and evolution of life (see Figure 1.6). Therefore, it is reasonable to have the number of Earth-like planets expected to typically orbit a "good" star as a parameter in the Drake Equation. For a planet to be characterized as Earth-like, it should not only have physical characteristics similar to the Earth (e.g., size, mass, composition, and geology) but also be in an orbit about its star that allows water to accumulate on its surface. As we shall discuss in later chapters, water is an essential ingredient for the origin and evolution of life, providing an ideal medium in which the chemical reactions necessary for life (as we know it) to occur. If a planet's orbit brings it too close to its star, the surface is too hot for water to exist in liquid form; it either boils or sublimates away (e.g., Mercury). Likewise, if it is too far from its star, water will have a tendency to freeze out on its surface (e.g., Mars). The range of orbital radii for which a planet's surface is neither too hot or too cold is referred to as the habitable zone of a planetary system. Other terminologies in the literature used to refer to this region include the "goldilocks zone" and "ecosphere". For liquid water to have an opportunity to accumulate on a planetary surface, the planet must have an atmosphere of sufficient pressure to slow the process of evaporation. Currently, within our solar system, there is only one such planet that meets all these criteria, *ours*. However, this may not have always been the case. Studies of both Mars and Venus show evidence for the presence

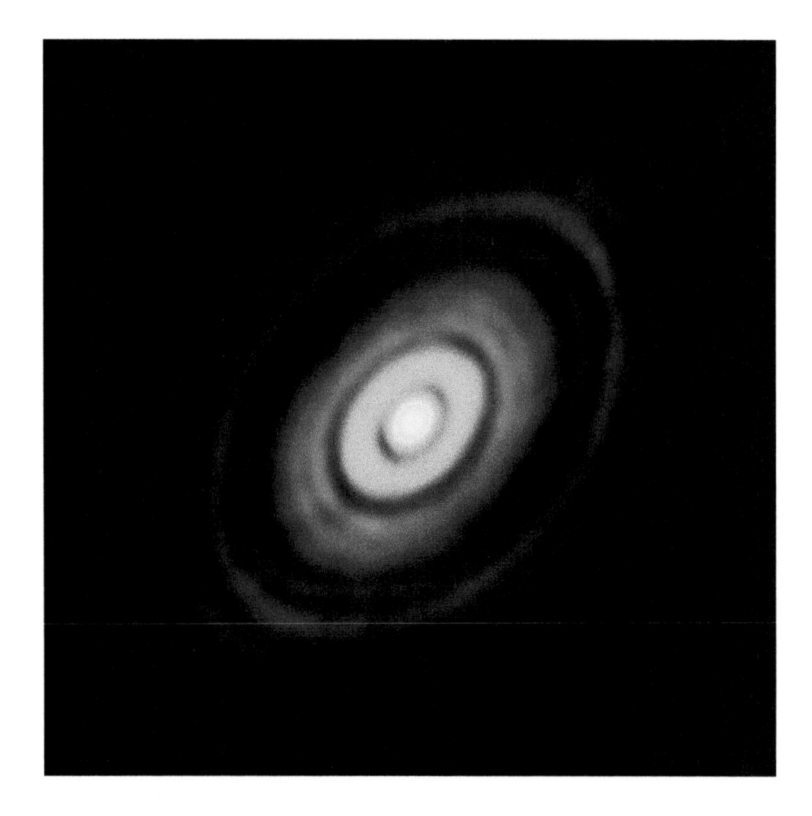

FIGURE 1.5 HL Tau Protoplanetary System. This image, made with the ALMA observatory, shows a planetary system similar to ours in the process of formation. The central object is a solar mass star surrounded by a protoplanetary disk. The dark rings may be due to the formation of protoplanets which have swept-up material within their orbital paths. Image Credit: NRAO, https://public.nrao.edu/gallery/hl-tau-birth-of-planets-revealed-in-astonishing-detail-2/

FIGURE 1.6 The Good Earth. Located within the Sun's habitable zone, our planet is the only world known to harbor life. However, observations suggest that conditions on both Mars and Venus may have been more amenable for the origin and evolution of life in the distant past. Image Credit: Shutterstock.

of liquid water on their surfaces for hundreds of millions of years after their formation. It is theorized their surface water was subsequently lost through a combination of solar, atmospheric, and geological processes. Whether this period of "Earth-like" conditions on these worlds was sufficient for the origin of life to occur is an open question that will be addressed by future planetary missions. For intelligent life to evolve on Earth required the presence of widespread surface water in the form of oceans and lakes for over ~95% of the Earth's ~4.7 billion year existence. What is the likelihood of finding another such planet around a Sun-like star? This is one of the primary questions driving exoplanet research. However, from combining Kepler's Laws of planetary motion with transit studies, we can now estimate that within the Milky Way, there could be more than 6 billion planets with a mass similar to the Earth orbiting within the habitable zones of Sun-like stars. While this is an impressively large number, this result suggests the value of n_e is less than 1. In other words, we will most likely need to survey a large number of planetary systems before we find another world that is truly Earth-like. For the sake of argument, let us assume we will need to survey a hundred Sun-like stars with planets before finding such a world, in which case n_e would be 0.01.

1.2.5 Fraction of Earth-like Planets where Life Originates: f_l

So far in our review of the Drake Equation, discoveries in astronomy and planetary science have been used to constrain its parameters. Recent advances in these fields have led to the detection and subsequent analysis of many planetary systems, thereby allowing their quantity and basic properties to be derived. For the next parameter, f_l, the fraction of Earth-like planets where life originates, the ability to make accurate estimates is more challenging for two reasons: (1) to date, no extraterrestrial life has been detected and (2) the time scale required to reproduce the origin of life in a laboratory setting may prove to be far greater than can be supported by human investigators. However, these bottlenecks can be overcome. First, we must continue to search for evidence of extraterrestrial life within our own solar system (and beyond), and second, continue to investigate and simulate the conditions under which life originated on the Earth, as well as the likelihood of those conditions occurring elsewhere. One of the first such attempts to recreate the origin of life by simulating conditions on primeval Earth was performed at the University of Chicago by Stanley Miller and Harold Urey in 1953. Using apparatus common to most chemistry laboratories, they were able to synthesize amino acids, the building blocks of life within a few days (see Chapter 6). This experiment has been successfully repeated and expanded upon many times since. Although no one has yet created organic life in the lab, these experiments suggest that the conditions under which life can originate have occurred on billions of worlds throughout the Milky Way. The fact that life evolved on Earth relatively soon after its formation suggests that if conditions are right, it may not be a rare occurrence (see Figure 1.7). While the likelihood of life originating on any one Earth-like planet may turn out to be small, i.e., f_l<<1, the sheer quantity of planetary incubators indicates the likelihood of finding extraterrestrial life of some form within the Milky Way is high. As a placeholder, let us assume there is a one in a hundred chance for life to originate on an Earth-like world, corresponding to an f_l of 0.01.

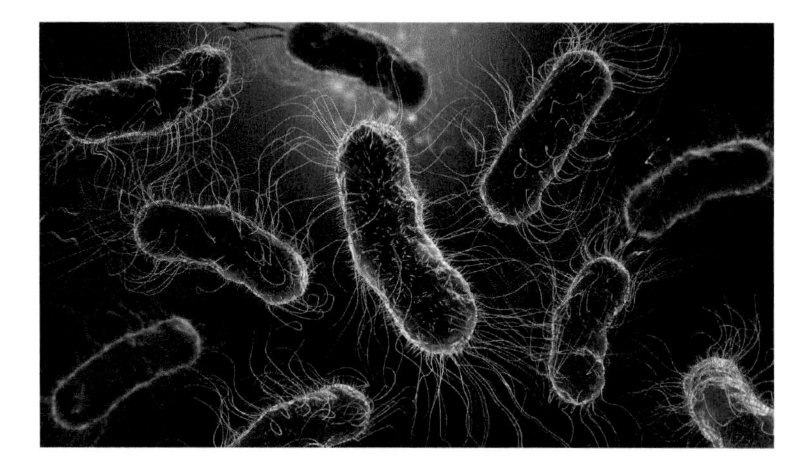

FIGURE 1.7 Halobacterium: an example of the type of simple, prokaryotic lifeform that could have been found on the primordial Earth. Image Credit: Shutterstock.

1.2.6 Fraction of Planets on which Intelligent Life Originates: f_i

For the purposes of the Drake Equation, an intelligence species can be defined as one having the cognitive ability to be aware of its own existence and of a universe that extends beyond the confines of its planet. Arguably, many creatures on Earth, across a variety of species, satisfy the first criteria, but, to our knowledge, only humans have looked up to the stars and contemplated their place among them. Physical evidence of this cosmic awareness can be seen in the form of cave paintings dating back to the Ice Age of the Upper Paleolithic, approximately 40,000 years ago (see Figure 1.8). The inkling of such awareness may date back much further, to at least the time of our *Homo erectus* ancestors, who lived between 1.5 and 1.9 million years ago. Even so, this is just the last blink of an eye

FIGURE 1.8 35,000–40,000-year-old cave painting in the Marcos-Pangkep karst (Sulawesi, Indonesia). Image Credit: Shutterstock.

in the 4.7 billion years of the Earth's existence. For much of this time, an alien survey ship visiting our planet would have found little more than simple, single-celled organisms in the Earth's lakes and oceans. It was not until the accumulation of free oxygen in the Earth's atmosphere and the subsequent catalytic effects it had on cellular chemistry, that more advanced, multicellular lifeforms emerged (see Chapter 7). From that point on, the evolution of more advanced, highly specialized lifeforms with ever-increasing cognitive abilities seems to have been inexorable. There were the occasional setbacks due to meteor and cometary impacts, geological events, weather conditions, or disease, but a continual march toward cosmic awareness occurred nonetheless. The history of life on Earth suggests that once the threshold of multicellular complexity in organisms is achieved, evolution is relentless in pushing toward intelligence. This bodes well for the likelihood of intelligence emerging in other worlds where life has achieved a foothold. Given this, let us assume a value for f_i of 0.3.

1.2.7 Fraction of Galactic Civilizations with the Desire and Ability to Communicate: f_c

Prehistoric art, as well as oral and written traditions, tells us humanity has looked toward the stars with the desire to communicate with intelligences greater than our own for thousands of years. For much of this time, the communication was spiritual in nature. Since the advent of agriculture about 10,000 years ago, cultural and technological evolution has brought humanity from the brink of extinction, with a worldwide population of less than 10 million, to where we are today, with an exponentially growing population of ~8 billion. After agriculture, arguably the greatest technological advance was writing, first on stone/clay tablets and later on papyrus and paper. This allowed knowledge to be more expeditiously shared, not only between coeval individuals but between generations as well. The dissemination of knowledge in the form of books increased dramatically with the invention of the printing press during the ninth century in China and the fifteenth century in Europe. Books allowed the hard-won knowledge of earlier generations to be passed on to the next, dramatically increasing their prospects for both survival and prosperity. Increased prosperity meant some fortunate individuals could commit significant fractions of their lives to basic research into what was once called "natural philosophy", which today we refer to as the natural sciences; e.g., math, physics, chemistry, biology, geology, and astronomy. Advances in the natural sciences are what has pulled back the veil of the unknown to provide us with insights into the nature of the universe. Engineering transforms discoveries in natural science into tools we can use to help advance our civilization. One of these advances is the ability to create and manipulate electromagnetic radiation for the purposes of communicating at the speed of light. This advance is commonly known as radio and is at the heart of every cell phone and television. Radio is a "gateway technology" for newly arrived species such as ourselves that affords the possibility of communicating between the stars at the speed of light without the need for starships (see Figure 1.9). Based on the human experience, for a civilization to survive and flourish, there must be both cultural and technological evolution (see Appendix 3: Timeline of Cultural/Technical Evolution). It is this evolutionary pressure that will work to drive an intelligent species to better understand its place in the universe and provide it with the tools needed

FIGURE 1.9 The NRAO Very Large Array (VLA) radio observatory in Socorro, NM. Image Credit: Shutterstock.

to communicate between the stars. However, there are natural and self-inflicted calamities that can slow or halt a species transition from an indigenous intelligence to an interstellar species. On Earth, the evolutionary clock has been reset several times (see Chapter 7). Based on this, let us assume a placeholder value for f_c of 0.1.

1.2.8 Lifetime of the Interstellar Communicative Phase

L_c: The human species officially passed the evolutionary threshold into the interstellar communicative phase on November 16, 1974. It was on this date that a high power, narrow band radio signal announcing our presence was beamed into space from the 1,000-ft Arecibo Radio Observatory in Puerto Rico. The digital message designed by Frank Drake and others contained information about our chemical make-up, physical size, and population, as well as the location of where the signal originated: the Earth (see Figure 1.10). This first effort lasted only 3 minutes and was only a symbolic gesture. The chances of an extraterrestrial civilization intercepting the signal are extremely remote. Nonetheless, it served as a demonstration of humanities' technological achievement and marked an important moment in the evolution of life on Earth. The next milestones in communicating our presence occurred in 2012 and 2018 when the Voyager 1 and Voyager 2 spacecrafts passed through the boundary of the heliosphere, the cocoon of charged particles and magnetic fields provided by the Sun and entered interstellar space (Colfield 2019). Each spacecraft carries a phonographic record conveying information about humanity and the location of the Earth.

At the time of this writing, the Arecibo Message is only 48 years old; this is the current value of L_c for humanity. This value of L_c is less than a single human lifetime. The time scales associated with evolution are typically measured in billions, millions, or thousands of years. What are the chances that another intelligent species has successfully emerged from the cauldron of evolution to be an interstellar communicative species in the past ~50

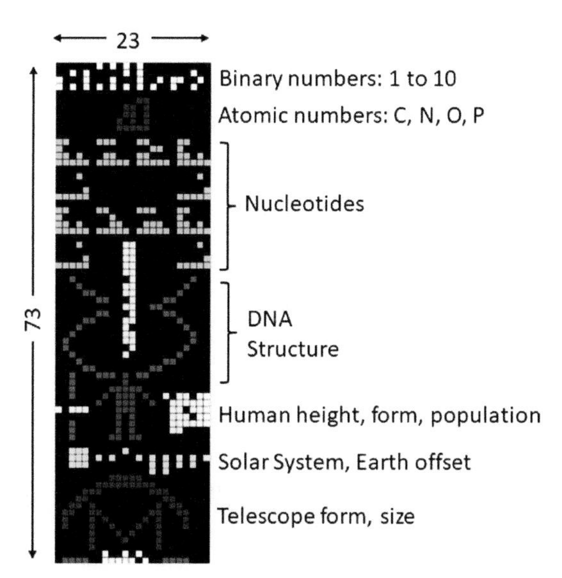

FIGURE 1.10 QR code for the Earth. Interstellar message sent from the Arecibo radio telescope on November 16, 1974. The message contained 23 × 73 bits (product of two primary numbers) and contains information describing life on Earth. The message was sent at a frequency of 2,380 MHz with an effective radiated power of ~18 terawatts. Image Credit: http://www.publicdomainfiles.com/show_file.php?id=13946131812000

years? The likelihood would seem very small. This implies that humanity may well be the youngest such species.

How long will our interstellar communicative phase last? The history of civilizations on Earth shows that their lifetimes can be subject to the occurrence of natural disasters that are out of their control. These include significant weather changes (e.g., ice ages or warming periods resulting in famine), disease (e.g., the bubonic plague), and geological events (e.g., volcanoes and earthquakes). A devastating war between rival factions can also end a civilization, as well as running out of the natural resources to sustain it. However, let us assume that cultural and technological evolution has (or soon will) advanced our species beyond the reach of these ancient enemies; what then will limit the lifetime of our civilization on Earth? The fossil and geological record indicates that meteor and/or cometary impacts have nearly wiped out life on Earth a number of times in the past. The last such major event occurred ~66 million years ago when an ~80-km-sized meteor struck the Earth near what is now the Yucatan Peninsula causing the extinction of most dinosaur species and paving the way for mammalian evolution, of which we are the result. Statistically, such large impacts are expected every ~50 million years, so we may be overdue. But, as the plot of several science fiction movies has suggested, it may soon be within our power to detect and deter the impact of meteors and comets in this size range. In which case, the lifetime of our civilization could be measured in millions of years. Ultimately, stellar evolution will cause the Sun to expand to the point where it will at first make the Earth uninhabitable and then, perhaps, engulf it. But this will not occur for several billion years. By then we will have likely long since left Earth and spread to the stars.

If we are pessimistic about the outcome of cultural evolution, then the value of L_c could be on the order of centuries. If, on the other hand, we are optimistic about the possible outcomes of cultural and technological evolution for us and other intelligent species, then L_c could be on the order of millions or billions of years. A method of estimating the minimum required lifetime of the communicative phase of neighboring civilizations is discussed in Appendix 5.

1.3 EVALUATING THE DRAKE EQUATION

Now that we have made "place-holder" estimates for each factor in the Drake Equation, we are in a position to perform an initial evaluation. Substitution of the values discussed above yields,

$$\begin{aligned} N_c &\simeq (4)(0.12)(0.7)(0.01)(0.01)(0.3)(0.1)L_c \\ &\simeq 1\times10^{-6}L_c \end{aligned} \tag{1.3}$$

The above result suggests that the likelihood of finding intelligent life in the Milky Way with which we can communicate is highly dependent on the average lifetime of the communicative phase of a species. If the pessimistic view of the likelihood of survival is correct, then $L_c \approx 100$ years and N_c is far less than one. This would suggest that while other civilizations may have once existed in the Milky Way, we are currently alone. If, however, a technically advanced civilization can overcome its baser instincts and become spacefaring, then L_c could be far greater. In which case, Eq. (1.3) would suggest we are part of a large galactic community with which we have yet to make contact.

1.4 HOW CLOSE ARE THEY?

By combining estimates of N_c with knowledge of the geometry and number density of stars in the Milky Way, an expression for determining the *average* distance to our nearest neighbors can be derived (Evans 1996). As discussed in Section 1.1, the Milky Way is a spiral galaxy, with a disk approximately 100,000 ly across and 1,000 ly thick. The number of star systems we would need to search, N_S, before we are likely to find an intelligent, communicative neighbor is then,

$$N_S = \frac{N_{MW}}{N_c} \tag{1.4}$$

where, N_{MW} is the number of stars in the Milky Way, estimated to be ~400 billion. If the neighbor is within a spherical "bubble" of space with a diameter < the thickness of the galactic disk, then Eq. (1.4) takes on the form,

$$N_S = n_* \left(\frac{4}{3}\pi r_S^3\right) \tag{1.5}$$

where,

$n_* =$ number density of stars in a cubic light year = 0.004 stars/ly^3

$r_S =$ search radius (ly)

If the search takes us out to star systems in the galactic disk beyond ~1,000 ly, then to first order we can approximate the galactic disk as a short cylinder, in which case Eq. (1.4) can take the form,

$$N_S = n_* \left(\pi h_d r_S^2\right) \tag{1.6}$$

where, $h_d =$ the height of the galactic disk ~1,000 ly.

Solving for the search radius, r_S, in each instance we find,

$$r_S = \begin{cases} \left(\dfrac{3}{4\pi}\dfrac{N_{MW}}{n_*}\right)^{\frac{1}{3}}\left(\dfrac{1}{N_c}\right)^{\frac{1}{3}} & \text{for } r_S \leq 1{,}000 \text{ ly} \\ \left(\dfrac{1}{\pi h_d}\dfrac{N_{MW}}{n_*}\right)^{\frac{1}{2}}\left(\dfrac{1}{N_c}\right)^{\frac{1}{2}} & \text{for } r_S > 1{,}000 \text{ ly} \end{cases} \tag{1.7}$$

A graphical representation of Eq. (1.7) indicating the average distance. r_S, between us and our closest interstellar neighbor as a function of N_c is shown in Figure 1.11.

1.5 SETI

Assuming the "Optimistic" case in Figure 1.11 is correct, then, on average, our nearest neighbor is ~500 ly away. In the near future (< 100 years), we will likely have developed the technology to build controlled nuclear fusion reactors, which can provide all the clean, safe energy our civilization will ever need. Soon after, we will likely use these same reactors to power our first manned interstellar spacecraft (Long 2011; Mahon 2018). These starships will be able to achieve velocities of 4–10% the speed of light. At this velocity, it would still take ~5,000 years to reach the nearest known Earth-like planet capable of supporting indigenous, intelligent life (see Section 8.4). Barring an unannounced visit (see the case for the interstellar object, "Oumuamua"; Bialy and Loeb 2018, Sutter 2022), radio is the best means of establishing communications between ourselves and our nearest neighbors. Being one of the youngest species capable of interstellar communications, it is then logical to assume that other more advanced civilizations may have anticipated our ascendance and are even now sending us welcome messages at radio frequencies (Cocconi and Morrison 1959). This is the idea behind SETI, the Search for Extraterrestrial Intelligence (Drake 2011; Tarter 2011). The first SETI survey was conducted at microwave frequencies in 1960 by Frank Drake (see Figure 1.12) and collaborators using an 85 ft diameter (26 m) radio telescope at the National Radio Astronomy Observatory in Greenbank, West Virginia (*Time* 1960). Since then, multinational SETI efforts have been underway. To date, there have been no

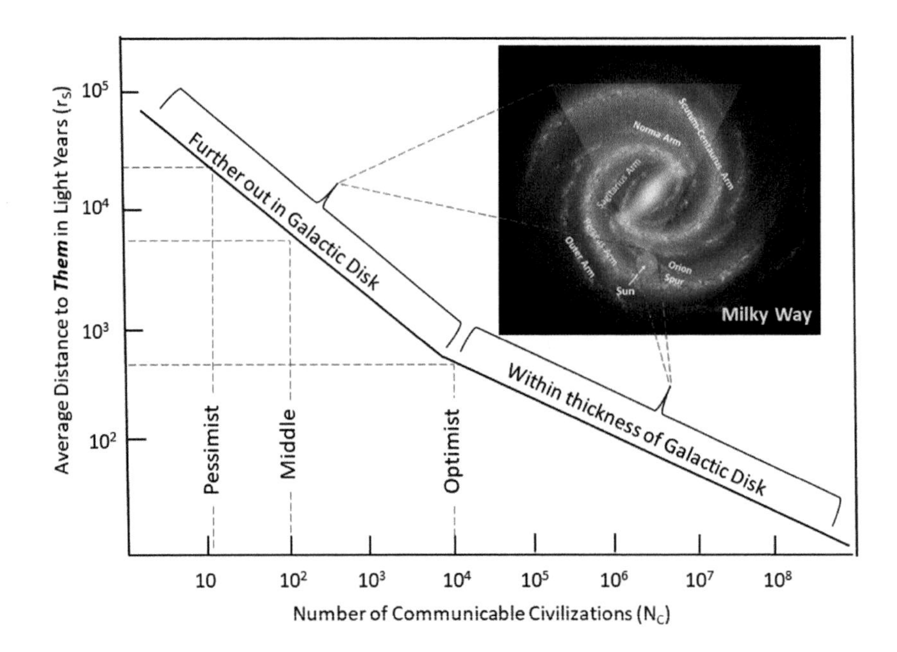

FIGURE 1.11 Average distance to nearest communicable civilization. The Drake Equation can be used to estimate the number of communicable civilizations, N_c, in the Milky Way. Using this estimate, together with our knowledge of the Milky Way's geometry and the number density of stars within it, the average distance between civilization, r_S, can be calculated. While recent data from space-based observatories suggest there could be ~5 billion Earth-like planets in the Milky Way, how many of these have (or will) nurture intelligent, communicable civilizations is unknown. The above figure illustrates three scenarios: "Pessimistic", "Middle", and "Optimistic", corresponding to the likelihood of such a civilization arising on an Earth-like planet being 2 in a billion, 2 in a 100 million, and 2 in a million, respectively. In the Optimistic scenario the distance to our nearest neighbor would be ~500 ly, while the Pessimistic case would place them ~24,000 ly away, roughly the distance between here and the Galactic Center. These are statistical averages, meaning our nearest neighbor could be by chance closer or further away. Image Credit: NASA/JPL-Caltech/ESO/R. Hurt.

confirmed detections. However, considering the number of potential targets and observing frequencies to choose from, SETI will likely continue as a multi-generational effort (see discussion in Appendix 5). Searches for waste heat from very advanced civilizations may be another means of detecting extraterrestrial life (Dyson 1960). However, a recent search in over 100,000 external galaxies did not yield an unambiguous detection (Griffith et al. 2015). In the near term (at least), it appears our best chance for finding life elsewhere will be in our searches of planetary bodies within our own solar system and the search for biosignatures in the atmospheres of exoplanets (see Chapter 8).

1.6 SUMMARY

In this chapter, we have introduced the Drake Equation and described how it can be used to estimate the number of communicable civilizations, N_c, within the Milky Way. We evaluated the Drake Equation using working estimates. We described how the estimate of N_c

FIGURE 1.12 Frank Drake ca. 1964. Radio astronomer who wrote the Drake Equation and performed the first Search for Extraterrestrial Intelligence (SETI). Image Credit: NRAO; used with kind permission of Frank Drake.

can be used together with knowledge of the distribution of stars within the Milky Way to estimate the distance, r_S, to our nearest interstellar neighbors. In the coming chapters, we will discuss the science and history behind the factors in the Drake Equation. In doing so, we will find that the investigation of life in the universe will reveal underlying relationships between the seemingly disparate fields of history, philosophy, mathematics, physics, astronomy, geology, chemistry, ecology, evolutionary biology, and engineering. We will discuss how this knowledge can be used to search for evidence of life in the solar system and beyond.

DISCUSSION QUESTIONS

1) Combining your own intuition with what you have learned in class and elsewhere, estimate each of the eight parameters on the right-hand side of the Drake Equation. Enter your estimates in the table below. Provide a brief explanation of how you arrived at your estimates. Include citations to any references (including web sites) you used.

R_*	f_g	f_p	n_e	f_l	f_i	f_c	L

2) Evaluate the Drake Equation and put your answer below.

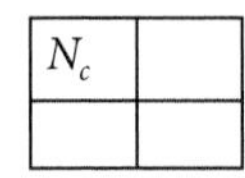

3) Assuming there are 4×10^{10} Sun-like stars in our galaxy and using your estimate for N_c, how many stars would you need to survey using a radio telescope before you are likely to find another communicable civilization?

4) If it only took 1 minute (i.e., 60 sec to examine each candidate star in your survey, how many years would it take to complete the survey?

5) Using Figure 1.11 and your estimate for N_c, how close would you estimate the nearest civilization to be?

6) Using radio waves, what would be the minimum time (in years) required for us to transmit a signal to and receive a signal back from the closest civilization?

REFERENCES

Bialy, S., and A. Loeb. 2018. "Could Solar Radiation Pressure Explain 'Oumuamua's Peculiar Acceleration?" *The Astrophysical Journal Letters* 868: L1.

Cocconi, G., and P. Morrison. 1959. "Searching for Interstellar Communications." *Nature* 184 (4690): 844–6.

Colfield, C. 2019. "Voyager 2 Illuminates Boundary of Interstellar Space." November. https://www.jpl.nasa.gov/news/voyager-2-illuminates-boundary-of-interstellar-space/.

Drake, F. 2011. "The Search for Extra-Terrestrial Intelligence." *Philosophical Transactions of the Royal Society of London Series. Part A* 369 (1936): 633–43.

Dyson, F. J. 1960. "Search for Artificial Sources of Infrared Radiation." *Science* 131 (3414): 1667.

Evans, N. 1996. *Extraterrestrial Life*. Edina: Burgess Publishing.

Griffith, Roger L., Jason T. Wright, Jessica Maldonado, Matthew S. Povich, Steinn Sigurđsson, and Brendan Mullan. 2015. "The Ĝ Infrared Search for Extraterrestrial Civilizations with Large Energy Supplies. III. The Reddest Extended Sources in WISE." *Astrophysical Journal: Supplement Series* 217: 25.

Long, K. F. 2011. *Deep Space Propulsion: A Roadmap to Interstellar Flight*. Springer, New York, NY.

Mahon, P. 2018. "Reaching the Stars in a Century Using Fusion Propulsion." August. https://i4is.org/reaching-the-stars-in-a-century-using-fusion-propulsion/.

Rix, H. W., and J. Bovy. 2013. "The Milky Way's Stellar Disk." *Astronomy & Astrophysics Review* 21 (1): 61.

"Science: Project Ozma." *Time*, April 18, 1960.

"SETI at 50." *Nature* 461 (7262) (2009): 316.

Sutter, P. 2022. "Will We Ever Know the True Nature of 'Oumuamua, the First Interstellar Visitor?'." *Space.com*, January 23. https://www.space.com/oumuamua-first-interstellar-visitor-true-nature-mystery.

Tarter, J. 2011. "Getting the World Actively Involved in SETI Searches". In *Communication with Extraterrestrial Intelligence*, edited by Douglas A. Vakoch, 71–80. State University of New York Press, Albany, NY.

CHAPTER 2

Setting the Stage for Cosmic Discovery

PROLOGUE

In the beginning, there was light. This light originated 13.77 billion years ago from a blast of immense power from a singularity occurring in multidimensional space–time. This event is referred to as the Big Bang. Before the Big Bang, the universe as we know did not exist. Just after the Big Bang, temperatures and densities were so high that only pure energy could exist. It was only after the universe expanded and cooled over several hundred thousand years that the first elementary particles (e.g., electrons and protons) could condense out of the light of the Big Bang. Shortly after the formation of these particles, the universe grew cold and dark. However, in the darkness, gravity was at work, drawing the rarefied protons and electrons together into gas clumps and filaments. Approximately 400 million years after the Big Bang, some of these clumps reached sufficient density to collapse under their own gravity to form the first stars. Once again, the universe contained visible light. At this time, the universe was still relatively compact, facilitating gravitational interaction between the primordial stars and interstellar gas clouds. In some instances, these interactions led to the assembly of gas clouds and stars into galaxies. The first generation of stars and interstellar clouds contained only the simplest of elements: hydrogen, helium, and some lithium. Heavier elements were forged in the centers of stars as byproducts of nuclear fusion. It is this nuclear fusion process that transforms the binding energy of atomic nuclei into the radiated energy that keeps a star from imploding under its own gravity and is the source of a star's light and heat. The centers of massive stars (~10× that of our Sun) are sufficiently hot and dense to forge all elements in the Periodic Table up to iron. Once a star reaches this point, the nuclear fire goes out, and it succumbs to gravity, imploding and often recoiling into a supernovae explosion. The immense energy and outward force of the supernovae create all the remaining naturally occurring elements and disperse the star's ash into space. The heavy elements contained in the stellar ash seeds the interstellar gas clouds from which a new generation of stars can be formed. After multiple generations

DOI: 10.1201/9781315210643-2

of star formation and supernova explosions, new stars became sufficiently rich in heavy elements that planets could form around them. It was on the surfaces (and perhaps interiors) of these planetary bodies that conditions were ripe for the origin and evolution of life. Many of the discoveries that have led to this theory of creation occurred over the last two centuries. In this chapter, we will give a brief account of some of the nineteenth-century individuals and innovations that helped pave the way to the cosmological discoveries of the twentieth.

2.1 INTRODUCTION

Until the advent and wide-scale use of the electric light bulb, on clear, moonless nights humans had a front-row seat to a pristine night sky. With the unaided eye they could see stars, planets, comets, gaseous nebula, and even dark dust clouds that are out of reach of the vast majority of modern humans. The night sky was not just the playground of the gods, it also served vital roles in the survival of ancient cultures by providing a means of telling time, marking the passing of the seasons, and navigation. For thousands of years, the human race has chronicled the goings on in the night sky using whatever medium was available. These renderings have been passed down to us in the form of cave paintings, petroglyphs, sculptures, mosaics, paintings, tapestries, and books (see Figure 2.1). For much of human history, the ability to observe celestial objects was limited by the size of the lens in the human eye (~10 mm). The apex in the ability to observe and record astronomical phenomena with the unaided eye was achieved by observers such as Tycho Brahe around 1600 AD. At almost the same time, the telescope was invented in Holland and subsequently used by Galileo Galilei for observing the heavens. Galileo's used a refractive telescope with a 37 mm diameter objective lens, yielding a light-collecting power ~13×

FIGURE 2.1 Petroglyph believed to be depicting the supernova of AD 1006. The supernovae are the star symbol right of center, with the constellation of Scorpio left of center. The boulder is located in White Tanks Regional Park, Phoenix, AZ. Image courtesy of John Barantine, Apache Point Observatory.

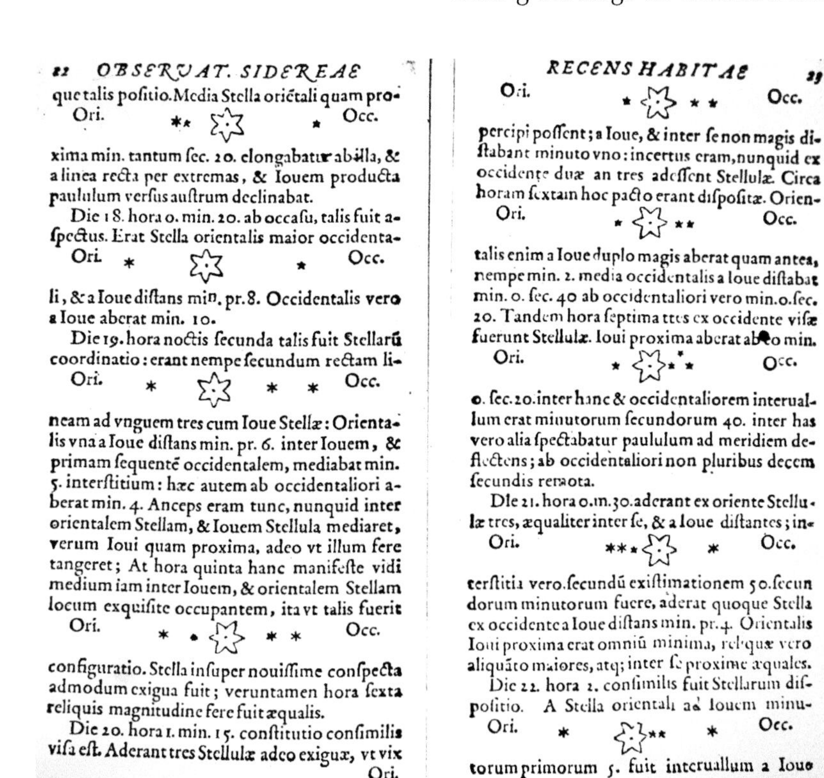

22 OBSERVAT. SIDEREAE

que talis poſitio. Media Stella oriẽtali quam pro-

Ori. Occ.

xima min. tantum ſec. 20. elongabatur ab illa, & a linea recta per extremas, & Iouem producta paululum verſus auſtrum declinabat.

Die 18. hora o. min. 20. ab occaſu, talis fuit aſpectus. Erat Stella orientalis maior occidenta-

Ori. Occ.

li, & a Ioue diſtans min. pr. 8. Occidentalis vero a Ioue aberat min. 10.

Die 19. hora noctis ſecunda talis fuit Stellarũ coordinatio: erant nempe ſecundum rectam li-

Ori. Occ.

neam ad vnguem tres cum Ioue Stellæ: Orientalis vna a Ioue diſtans min. pr. 6. inter Iouem, & primam ſequentẽ occidentalem, mediabat min. 5. interſtitium: hæc autem ab occidentaliori aberat min. 4. Anceps eram tunc, nunquid inter orientalem Stellam, & Iouem Stellula mediaret, verum Ioui quam proxima, adeo vt illum fere tangeret; At hora quinta hanc manifeſte vidi medium iam inter Iouem, & orientalem Stellam locum exquiſite occupantem, ita vt talis fuerit

Ori. Occ.

configuratio. Stella inſuper nouiſſime conſpecta admodum exigua fuit; veruntamen hora ſexta reliquis magnitudine fere fuit æqualis.

Die 20. hora 1. min. 15. conſtitutio conſimilis viſa eſt. Aderant tres Stellulæ adeo exiguæ, vt vix

RECENS HABITAE 23

Ori. Occ.

percipi poſſent; a Ioue, & inter ſe non magis diſtabant minuto vno: incertus eram, nunquid ex occidente duæ an tres adeſſent Stellulæ. Circa horam ſextam hoc pacto erant diſpoſitæ. Orien-

Ori. Occ.

talis enim a Ioue duplo magis aberat quam antea, nempe min. 2. media occidentalis a Ioue diſtabat min. 0. ſec. 40 ab occidentaliori vero min. 0. ſec. 20. Tandem hora ſeptima tres ex occidente viſæ fuerunt Stellulæ. Ioui proxima aberat ab eo min.

Ori. Occ.

0. ſec. 20. inter hanc & occidentaliorem interuallum erat minutorum ſecundorum 40. inter has vero alia ſpectabatur paululum ad meridiem deflectens; ab occidentaliori non pluribus decem ſecundis remota.

Die 21. hora 0. m. 30. aderant ex oriente Stellulæ tres, æqualiter inter ſe, & a Ioue diſtantes; in-

Ori. Occ.

terſtitia vero ſecundũ exiſtimationem 50. ſecundorum minutorum fuere, aderat quoque Stella ex occidente a Ioue diſtans min. pr. 4. Orientalis Ioui proxima erat omniũ minima, reliquæ vero aliquãto maiores, atq; inter ſe proxime æquales.

Die 22. hora 2. conſimilis fuit Stellarum diſpoſitio. A Stella orientali ad Iouem minu-

Ori. Occ.

torum primorum 5. fuit interuallum a Ioue

FIGURE 2.2 A copy of Galileo's observations of Jupiter and its satellites from the seventeenth century. Image Credit: Alamy.

greater than the human eye. With it, he could see farther and make out details of celestial objects that were theretofore unknown. Discoveries made by Galileo with his homebuilt telescope include the four moons of Jupiter, the fact that Venus goes through phases like the Moon (thereby disproving Ptolemy's geocentric model of the solar system), the planet Neptune, sunspots, mountains on the Moon, and the multitude of stars associated with the Milky Way. Each of these observations broke new ground in our understanding of the universe. Galileo sketched each one as faithfully as he could in his observing notebook (see Figure 2.2). While the sketches served their purpose, they were subject to artistic expression and limited in quantity by the speed at which Galileo could draw them. Over the next two centuries, the size of telescopes increased by a factor of ~100, leading to an increase in light-gathering power of ~10,000. However, the ability to faithfully record what was observed remained limited by the ability of astronomers to sketch what they saw. The invention that broke this bottleneck and enabled astronomers to realize the full potential of their telescopes was photography.

2.2 ORIGIN OF PHOTOGRAPHY

Photography is the art or practice of creating durable images by recording light on a medium, e.g., paper, film, or electronically. This is achieved by using a camera, the Latin word for chamber. The first written description of a camera is found in the fourth-century BC Chinese text called "Mozi". The text describes how the inverted image of an illuminated

FIGURE 2.3 The Camera Obscura. Known since ancient times, the camera obscura consists of a darkened chamber with a pinhole in one wall. Light enters the chamber through the pinhole and projects an inverted image of the outside scene on the opposing wall. Until the invention of photography, there was no way to permanently record the image other than having the observer trace it on cloth or paper. Image Credit: Shutterstock.

person outside a darkened chamber can be cast on the far wall of the chamber through a small hole in the wall closest to the person (see Figure 2.3). Some years later the Greek philosopher Aristotle (384–322 BC) described a similar phenomenon in his work, "Problems", Book 15. A detailed physical optics description of the phenomenon can be found in the "Book of Optics" by the Arab physicist Ibn al-Haytham (ca. 1021 AD). Later authors would refer to such a device as a "pinhole camera" or "camera obscura", Latin for "dark chamber". Camera obscuras utilizing a lens in the opening to increase light-gathering power (and therefore image brightness) have been used by artists since the late 1500s to project images on to surfaces, thereby enabling them to trace or paint scenes with greater fidelity than could be achieved free-hand or from memory. This would be done with the artist literally inside the camera, i.e., a darkened room. If the wall on which the scene was projected was translucent, for example, a thin sheet of paper, the artist could trace the image from the other side, avoiding blocking the incoming light. Some famous painters who may have employed some form of the camera obscura in creating their art include da Vinci, Rembrandt, Caravaggio, and Vermeer (Hockney-Falco theory; Stork 2004; Dupre 2005). The first portable camera obscura was developed by Johannes Kepler in 1604, who coined the term, "camera obscura". Incidentally, five years later (after working with Tycho Brahe), Kepler introduced the first two laws of planetary motion (more on that later). The next breakthrough in camera technology came ~200 years later with the development of chemical photography by Nicephore Niepce.

2.3 NICEPHORE NIEPCE

As is often the case with inventors, Niepce (see Figure 2.4) came from a diverse background. He was born into a wealthy family in Chalon-su-Saone, France in 1765 and came of age during the French Revolution, in which he served as a staff officer under Napoleon. He resigned his commission due to poor health and became the administrator of the district of Nice. Before joining the army, he studied science and the experimental method at the Oration College of Angers. In 1795, he resigned his administrator position and returned home to the family estate in Chalon to pursue his interest in scientific research with his older brother Claude.

2.3.1 The Pyreolophore

One of the inventions they worked on was the world's first operating internal combustion engine, which they called the "Pyreolophore" (a combination of the Greek words for "fire", "wind", and "I produce"). The theoretical design for such an engine was discussed more than a hundred years earlier (1680) by the Dutch scientist/astronomer Christiaan Huygens, but it was not until 1806 that a working prototype was created by the Niepce brothers. The next year they received a 10-year patent for their invention, which was signed by Emperor Napoleon Bonaparte. A diagram of the inner workings of their engine is shown in Figure 2.5. It has a cylinder, piston, combustion chamber, timing mechanisms, and valves,

FIGURE 2.4 Joseph Nicephore Niepce (ca. 1795), inventor of photography and co-inventor (with his brother Claude) of the internal combustion engine. Image Credit: Alamy.

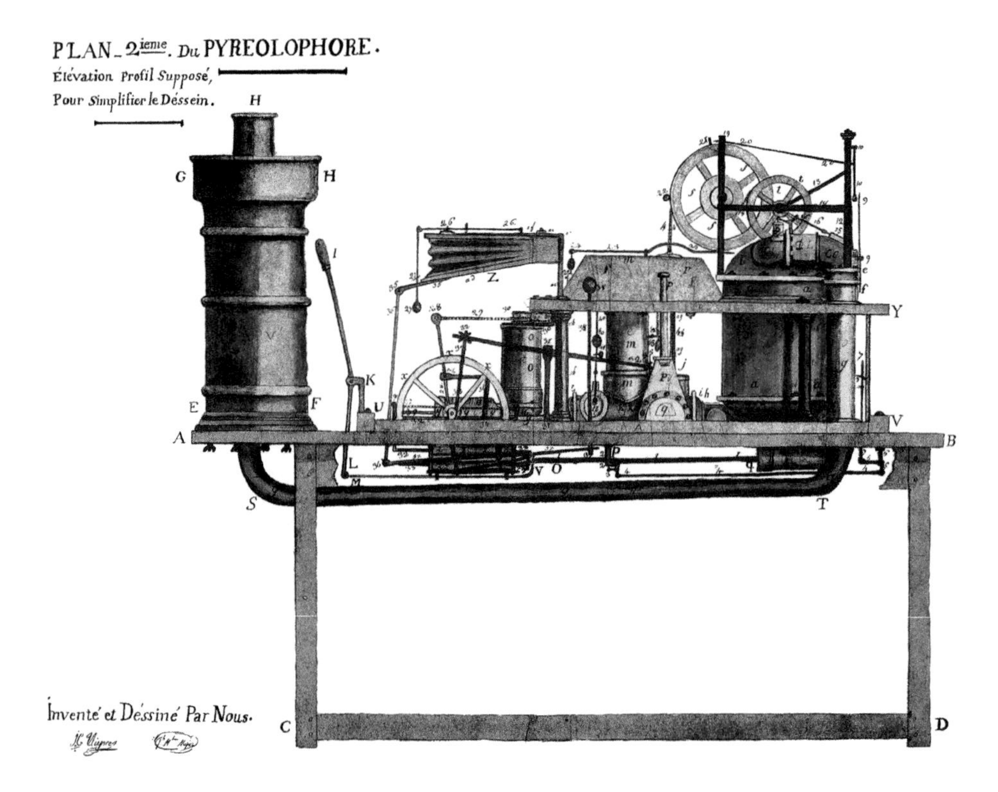

FIGURE 2.5 Niepce brothers' plan for the first internal combustion engine ca. 1807. Image Credit: Alamy.

much like modern engines. Early models used combustible powders of various sorts (e.g., Lycopodium, coal, together with resin). Later models used white oil of petroleum. The engine was first put to use by the brothers to power a one-ton boat up the river Saone (Maison Niepce 2020). This first engine ran at 12–13 rpm (revolutions per minute) with sufficient power to drive the boat upstream. The engine did not turn a propeller, but instead, sucked water in the front of the boat and pushed it out the back. Over the 10-year period of the French patent, they made a number of improvements to the design concept (including fuel injection) but were unable to make it a commercially viable product. Claude traveled to London and was able to file a new patent, this time signed by King George III. Shortly thereafter (1818), Nicephore built one of the first bicycles in France and wrote to Claude suggesting they adapt their engine to power it, which would have been the first motorcycle. Claude spent the next 10 years in England promoting their engine, but once more was unsuccessful in securing the funding needed to make it commercially viable. In the process, Claude spent most of the family fortune and, while suffering from depression, descended into delirium. He died in 1828, soon after a visit from his brother. While the brothers were the first to bring the internal combustion engine into reality, it was not until 1860 that a Belgium, Etienne Lenoir, made the first commercially successful model. It used a much more energy-efficient fuel, coal gas, and employed an electrical ignition system. However, in the 10-year period leading up to his brother's death, Nicephore had been hard at work on another Niepce family invention, photography.

2.3.2 First Photographs

One of Nicephore Niepce's interests was the recently invented artform of lithography, from the Greek word for "lithos" meaning stone, whereby an image etched in stone could be repeatedly printed onto paper. In the process invented in 1798 by Alois Senefelder, an artist would first draw a scene on a porous surface, typically limestone, with a grease pen. The limestone is then dipped in an acidic solution (e.g., vinegar). The surface area not protected by the grease is etched, leaving the artist's scene elevated relative to the rest of the stone's surface. To create a print, the artist would first wet the surface of the stone with water. The greased image repels the water. A roller is then used to apply ink to the stone, with the ink only adhering to the greased image. A sheet of dry paper is then pressed onto the surface, thereby transferring the inked image to the paper. Another approach was to use Bitumen of Judea, a naturally occurring acid-resistant asphalt, to coat a copper plate. The artist would then scratch their drawing through the coating. The plate is then placed in an acid bath to etch the unprotected, scratched areas. Next, the plate is rolled with ink. The ink fills the etched traces. As before, by pressing paper on the inked plate, the image is transferred to paper. Niepce found that if the bitumen was dissolved in lavender oil and then applied in a thin layer on to the lithographic stone (or a sheet of copper or glass), it became harder when exposed to light, with a degree of hardness proportional to the intensity of the light cast upon it. He used this discovery to make the first "photocopy" by laying a paper with an image on it over a plate thinly coated with bitumen and exposing it to sunlight. The lines or dark areas on the paper would block the light from reaching the underlying bitumen-coated plate, leaving those areas on the plate susceptible to an acid wash, just like the regions that were hand-etched by the artist in the original process. Niepce called this process heliography, an example of which is shown in Figure 2.6. This image was produced in 1825. His basic concept for photocopying is still in use today and employed in

FIGURE 2.6 Heliograph on an ink-on-paper print made by Nicephore Niepce in 1825. The demonstration of utilizing a light-sensitive emulsion to create images was a critical step along the path to the invention of photography. Image Credit: Alamy.

everything from making copies of office memos to making computer chips using contact photolithography.

As discussed earlier, artists at the time were known to trace "light paintings" cast on paper by a camera obscura to create artwork. Niepce's breakthrough idea was to substitute a photosensitive, bitumen-covered plate he had developed for lithography for the tracing paper and see if he could directly capture the projected image. Based upon correspondence with his brother Claude, we know Niepce first succeeded in this endeavor in 1824 using a camera obscura and a bitumen-covered lithographic stone. This original image was lost, but Niepce repeated the experiment in 1826 or 1827 using a sheet of bitumen-coated pewter. The image survives to this day and is shown in Figure 2.7. Due to the limited sensitivity of his coatings, the exposure lasted eight hours. Niepce called the process heliography, since it required bright sunlight to work. The image is the view from his house. A few years later Niepce sought to build upon his success by partnering with Louis Daguerre, a Parisian artist, also interested in creating permanent photographic images. They improved upon the process by using lavender oil distillate instead of bitumen as the photosensitive material. However, the cost of further developing the process led to Niepce's financial ruin. Upon his death in 1833, his family was unable to pay for a funeral, and he was laid to rest in a pauper's grave. Over the next six years, Daguerre continued improving on the sensitivity of the process by utilizing iodized silvered plates developed in Mercury vapors and "fixed" with sodium thiosulfate. The required exposure times were 60–80 times shorter than with bitumen, making photography both a powerful tool for recording events and commercially viable. The work culminated in what Daguerre called the "daguerreotype". The first daguerreotype with a recognizable human was taken in Paris by Daguerre in 1838

FIGURE 2.7 Oldest surviving photographic image. Taken by Niepce using a camera obscura and bitumen-covered pewter plate from his house, ca. 1827 (Public Domain). The lighting suggests the exposure was performed over several days (Niepce House Museum 2013).

FIGURE 2.8 First photograph with recognizable humans. This image was taken by Louis Daguerre, a former partner of Niepce, in 1838 using the process that bears his name. The human beings are in the lower left-hand corner, where a man is getting his shoes shined. The image was taken on a busy street in Paris. Due to the relatively long exposure times required by this process, everyone and everything else was moving too fast to be recorded. Image: Public Domain.

(see Figure 2.8). A picture of a daguerreotype camera is shown in Figure 2.9. Many of the nineteenth-century images of historical figures you may have seen are daguerreotypes. Ultimately, the French government purchased the invention and awarded both Daguerre and the Niepce family yearly stipends for their groundbreaking work in what is now referred to as photography (Gorman 2007).

Once the arrangement for royalty payments was secured, Daguerre gave a step-by-step description of the process before a joint session of the Academie des Sciences and the Academie des Beaux-Arts on August 19, 1839. As part of the deal, Daguerre was required to publish detailed instructions for the process, which quickly made it across the Atlantic and into the hands of Robert Cornelius, an amateur chemist and photography enthusiast. By October of the same year, he had prepared photographic plates and set-up a camera in the back of his father's gas-light business on Chestnut Street in Philadelphia. After removing the lens cap, he ran and sat very still in front of the camera for over a minute before replacing the cap. The picture he developed is shown in Figure 2.10 and believed to be the first "selfie". On the back of the picture, Cornelius wrote, "The first Light picture ever taken. 1839" (Petapixel 2011). The first portrait of a female subject was taken at about the same time by John William Draper, a professor of chemistry at New York University. He first photographed his assistant using a 65-sec exposure in sunlight. Unfortunately, that image did not survive. However, a contemporaneous photograph of Draper's sister, taken in late 1839, survives is shown in Figure 2.11 (McManus 1995). Shortly after (March 26, 1840), Draper had significantly improved the sensitivity of the Daguerre's process such that he was able to use the rooftop observatory at New York University to make the first photograph of the Moon (see Figure 2.12).

FIGURE 2.9 Daguerreotype camera constructed by Maison Susse Freres, ca. 1839. Image Credit: Shutterstock.

FIGURE 2.10 First Portrait of a Human. This "selfie" was taken by an amateur chemist, Robert Cornelius, outside the back of his father's store in Philadelphia on a sunny day in October of 1839. The exposure time was ~1 min. Image: Public Domain.

FIGURE 2.11 First portrait of a Female. The subject, Dorothy Catherine Draper, was the sister of the photographer, John Draper, a chemistry professor at New York University. The image was likely taken in late 1839. Image: Public Domain.

2.4 THE FIRST SPECTROGRAM

As was demonstrated by Isaac Newton and others since antiquity, when the light of the Sun is passed through a glass prism, refraction effects in the prism will disperse the light beam into a rainbow of colors. As light passes through a prism, the paths of the shorter-wavelength (i.e., blue) components of the light are bent more than the longer-wavelength (i.e., red) components. Water droplets in the atmosphere can act like small prisms, both refracting and reflecting sunlight such that one or more rainbows are perceived by an observer. The wavelength dependence of the refraction is what causes the outer arc of the primary (i.e., the brightest) rainbow to appear red and the inner arc blue. Additional reflections within raindrops can lead to the appearance of secondary rainbows where the order of colors is progressively reversed. A rainbow is, in fact, the optical spectrum of the Sun. A spectrum is simply a plot of the observed intensity (or brightness) of an object versus wavelength (or frequency). In 1802 the English scientist Willian Hyde Wollaston built an improved spectrometer that included a lens to focus the Sun's spectrum on a screen. With it, he observed that the Sun's spectrum was not continuous, but some ranges of colors were missing, appearing as black bands. These dark features became known as "absorption lines" (see Figure 2.13 *left*). Wollaston mistakenly thought these features corresponded to

FIGURE 2.12 First photograph of the Moon. This Daguerreotype was taken by John William Draper from the roof of New York University in 1840. Image: Public Domain.

FIGURE 2.13 *Left:* Modern Solar Spectrum showing Fraunhofer Absorption Lines. *Right*: First daguerreotype of a solar spectrum taken by John W. Draper in 1843. Draper's notes and signature are shown. This is a negative, so the dark absorption features appear white. Images: Public Domain.

naturally occurring breaks between colors. Later, in 1814 a German physicist, Fraunhofer, rediscovered these absorption lines and began a systematic study of them, labeling the darkest and widest ones with letters A through K. He used other letters for the weaker ones. The sheer number of the lines he observed (~600) ruled out Wollaston's boundary hypothesis. In 1843, John Draper made the first daguerreotypes of the solar spectrum, identifying not only the Fraunhofer lines but also infrared and ultraviolet absorption features outside the tuning range of the human eye (Daniel 1938). The daguerreotype of the solar spectrum taken by Draper is shown in Figure 2.13, *right*.

2.5 KIRCHHOFF AND BUNSEN

In 1859 two German physicists, Gustav Kirchhoff and Robert Bunsen (see Figure 2.14), demonstrated that spectral absorption and emission lines can be used to identify elements within the medium from which they are observed. In 1859 Kirchhoff showed the same spectral pattern observed in the lab for the element sodium is also observed in the Sun (the absorption line labeled "D" in Figure 2.12), thereby providing proof of its existence in the

FIGURE 2.14 Gustav Kirchhoff (left) and Robert Bunsen (right). Together they showed how spectroscopy can be used to remotely identify substances in distant objects (i.e., the Sun). Image: Public Domain.

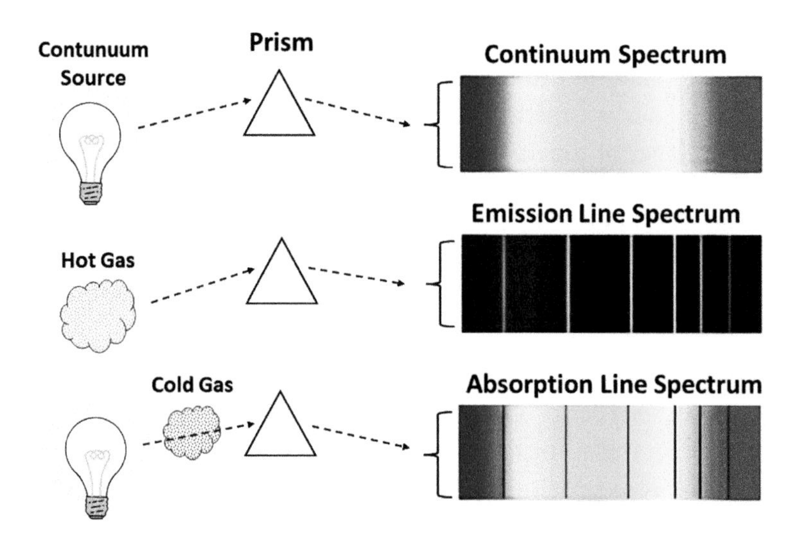

FIGURE 2.15 Kirchhoff's 3 Laws of Spectroscopy. When hot dense objects (e.g., the filament of a lightbulb) are observed through a spectroscope (e.g., a prism), a continuous spectrum is seen. A low-density, hot gas yields an emission line spectrum. In the case of a hot dense object with a colder gas in front, an absorption line spectrum (e.g., Fig. 12) is observed.

Sun's atmosphere (Kirchhoff 1859). From this work, Kirchhoff formulated his three Laws of Spectroscopy (see Figure 2.15).

1. A solid, liquid, or dense gas heated to emit light will do so at all wavelengths, thereby producing a continuous spectrum (like an uninterrupted rainbow).
2. A rarefied (i.e., low-density) gas when heated will emit light only at specific wavelengths, thereby producing an emission line spectrum.
3. If light from an object with a continuous spectrum passes through a cooler, low-density foreground gas, an absorption spectrum will be produced. The wavelengths of the absorption lines correspond to the wavelengths of the emission lines that would be observed from the gas alone.

2.6 HUGGINSES AND DRAPERS

The news of Kirchhoff's discovery of sodium in the Sun soon reached the amateur British astronomer William Huggins, who set about to build a spectroscope that could be used on his telescope to determine the composition of astronomical objects. Together with William Allen Miller, professor of chemistry at King's College, they built a spectrograph with which they could directly compare the emission spectra of a heated sample of material (e.g., hydrogen, sodium, and iron) to the spectra of the astronomical object under study. Using an 8" Alvan Clark refracting telescope, their initial work focused on the observation of solar system objects, with a first publication in 1862. One of his greatest discoveries was made on the evening of August 29, 1864, when he observed a single bright emission line on the output of his spectrograph when his telescope was pointed toward a planetary nebula in the constellation of Draco. Kirchhoff's 2nd Law of Spectroscopy dictated that such an emission line spectrum could only originate in hot gas. This proved that such nebulae are not an aggregation of stars but a luminous gas. This was the first definitive observational proof of the existence of a gaseous component of the interstellar medium. By 1868 Huggins had begun observing the spectra from distant stars and noted a displacement in the spectral location of emission lines toward the star Sirius compared to his local reference spectra. He concluded the shift was due to the Doppler effect, indicating the star was moving at a velocity of over 32 km/s relative to the Earth (Huggins 1868). As we will learn, the ability to use spectroscopy to determine not only the composition but also the relative velocity of astronomical objects will prove key to our understanding of the structure and evolution of the universe. Huggins first attempted to record spectra photographically in 1863, something William Draper had done 20 years earlier, but he found the use of wet photographic plates with his instrument to be impractical.

However, across the Atlantic, Dr. Henry Draper (see Figure 2.16), the son of Dr. John William Draper, was carrying on his father's work in astrophotography and in 1862 completed the construction of a powerful, state-of-the-art, reflecting telescope with a 15.5-inch diameter silvered glass mirror. Shortly before this time, telescope mirrors were made from polished metal (i.e., speculum). Glass-based mirrors were much lighter and easier to polish, thereby making it possible to make larger and more accurate mirrors capable of yielding higher-quality images. After spending a year as a surgeon during the American Civil War, he returned to his homemade observatory and took 1500 photographs of solar system objects and began spectroscopic studies. In 1867 he married Anna Palmer (see Figure 2.16, *right*), and during their honeymoon, he selected the glass that would become the mirror of a new 28-inch telescope. It took 5 years of effort to build the new telescope. During this time, they used their existing telescope to make the best spectrograms of the Sun ever made and discovered oxygen in the solar atmosphere. On completion of the 28-inch telescope in August 1872, they succeeded where Huggins had failed and made their first photograph of a stellar spectrum. They observed four hydrogen lines in the light of Vega. Meanwhile, between 1875 and 1876, Huggins had successfully switched to using rapid dry, silver bromide photographic plates to record spectrograms. The successful application of this technical approach is contemporaneous with his marriage to accomplished photographer and amateur scientist Margaret Lindsay Murray.

In the spring of 1879, the Drapers traveled to England and met with the Huggins at their home observatory in Tulse Hill. On their return to New York, the Drapers took the Huggins' advice and switched to dry photographic plates. A few years later, the Drapers took the photograph of the Orion Nebula shown in Figure 2.17. This image demonstrated that with sufficient exposure time, it was possible to make photographic images with sensitivities exceeding that of the human eye. With the improved sensitivity, the Drapers began to perform spectroscopic surveys of distance stars, a few of which are shown in Figure 2.18. They began classifying them by spectral type, noting the spectra's similarities and differences compared to the Sun. Unfortunately, soon after these spectra were taken, Henry Draper developed pleurisy after a hunting trip in the Rocky Mountains and died of pneumonia in November 1882. However, the Draper

FIGURE 2.16 Left: Henry Draper at his telescope, ca. 1862. *Right:* Anna Palmer Draper, ca. 1867. Images: Public Domain.

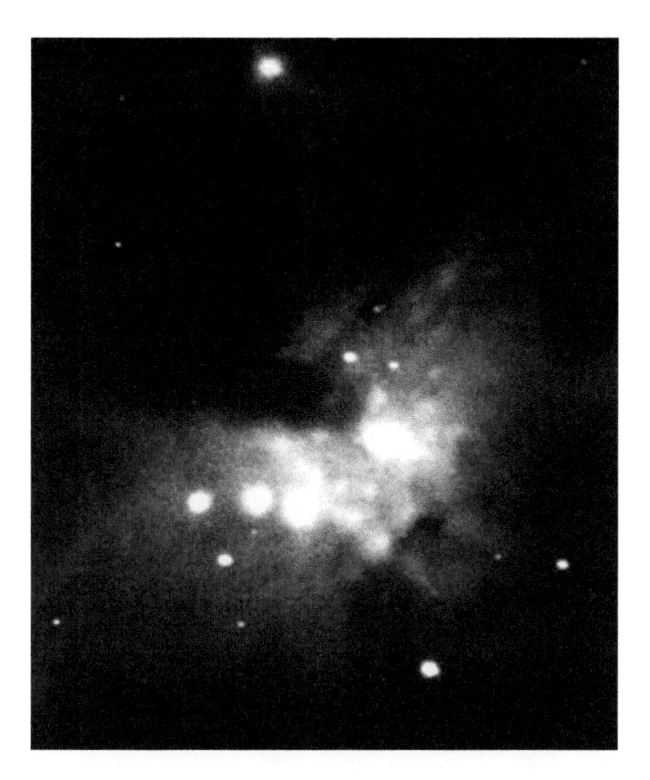

FIGURE 2.17 Image of the Orion Nebula taken by the Drapers, ca. 1880. Image: Public Domain.

story does not end here. While hosting a National Academy of Science conference banquet at their home just before Henry's death, the Drapers met Edward Pickering, the director of the Harvard College Observatory who was very interested in their work on stellar classification. After her husband's untimely death at age 45, Anna Draper donated their telescopes and a sizable endowment (the Henry Draper Memorial) to the Harvard College Observatory for the purpose of continuing their work on stellar classification. Pickering put the donations to good use, hiring a gifted group of astronomers, mostly female, for this momentous task. The first publication from this work was in 1890, "The Draper Catalogue of Stellar Spectra" (Tenn 1985), which contained the spectral classification of ~10,000 stars. Meanwhile, back in England, for the next 30 years, the Huggins (see Figure 2.19) devoted themselves to improving their astronomical instrumentation and using it to advance our understanding of the chemical composition of astronomical objects. Together they published the "Atlas of Representative Stellar Spectra" in March 1900 (Huggins 1899; Campbell 1900).

2.7 SUMMARY

For most of the history of humankind the only way to convey images of the night sky from one generation to another were through hand-drawn illustrations; first on stones or cave walls and later on papyrus, cloth, and paper. Knowledge of the detailed structure of astrophysical objects improved with the advent of the telescope in the seventeenth century. However, it was not until the nineteenth century that the inventiveness and steadfast work

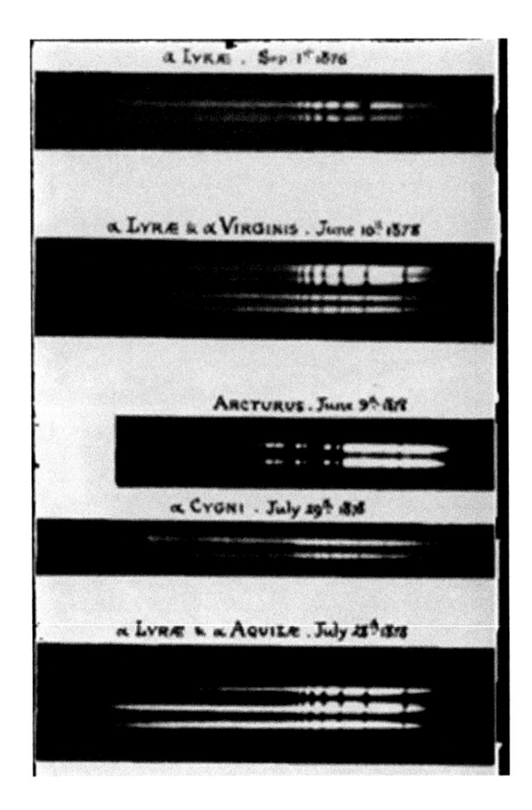

FIGURE 2.18 Stellar spectra taken by the Drapers in 1882. (Atlas of Representative Stellar Spectra, 1889.) Images: Public Domain.

FIGURE 2.19 *Left*: William Huggins (1824–1910); *Right*: Margaret Huggins (1848–1915). This husband and wife team helped pioneer the application of high-resolution spectroscopy to the study of planets, stars, nebula, and galaxies. Margaret is pictured at the Tulse Hill Observatory next to the high-resolution spectrograph they developed and used for many of their discoveries. Their work, together with that of the Drapers, opened the door to modern astrophysics, allowing us to better understand the physical properties (e.g., composition and dynamics) of objects throughout the observable universe. Images: Public Domain.

of early pioneers in the areas of photography and spectroscopy provided the diagnostic tools needed to open the door to a new era of cosmic discovery that was soon to come.

REVIEW QUESTIONS

1) Before the telescope was invented, what was the largest lens used to observe astronomical phenomena (or anything else for that matter)? Where did it reside?
2) Galileo made many discoveries when he first used a telescope to perform studies of the night sky. Which one do you find the most interesting and why?
3) What ancient apparatus is the design of the camera based on? How did it work?
4) Who took the first photograph? Why was the exposure time so long?
5) What other inventions did the creator of photography make? What was his reward?
6) Where are all the people in Figure 2.8?
7) Why was the invention of photography important for the advancement of astronomy?
8) When was the first selfie made?
9) Who made the first photograph of an astronomical object? What was it made of?
10) What is a Fraunhofer spectrum?
11) Who made the first spectrogram of an astronomical object? What did he find?
12) What are Kirchhoff's 3 Laws of Spectroscopy? What was their first application in astronomy?
13) Who first measured a Doppler shift in an astronomical object and an emission line in a nebula?
14) Which power couple made the first spectrogram of a distant star?
15) Where did the Harvard College Observatory initially get its funding to conduct a spectroscopic survey of stars?

REFERENCES

Campbell, W. M. 1900. "An Atlas of Representative Stellar Spectra: A Review." *Publication of the Astronomical Society of the Pacific* 12 (76): 246.

Dupre, S., 2007. "Playing with Images in a Dark Room Kepler's *Ludi* inside the Camera Obscura" in *Inside the Camera Obscura – Optics and Art under the Spell of the Projected Image*, ed. Wolfgang Lefevre, pub. Max Planck Institute for the History of Science, Berlin, Germany, p. 59.

Gorman, Jessica. 2007. "Photography at a Crossroads." *Science News* 162 (21): 331–3. https://doi.org/10.2307/4013861.

Huggins, Sir William and Lady. 1899. *An Atlas of Representative Stellar Spectra*. London: William Wesley and Son, 6–7.

Huggins, W. 1868. "Further Observations on the Spectra of Some of the Stars and Nebulae, with an Attempt to Determine Therefrom Whether These Bodies Are Moving Towards or from the Earth, Also Observations on the Spectra of the Sun and of Comet II." Philosophical Transactions of the Royal Society of London 158: 529–64.

Kirchhoff, G. 1859. "Presentation to the Berlin Academy." October 27.

McManus, Howard R. 1995. "The Most Famous Daguerreian Portrait: Exploring the History of the Dorothy Catherine Draper Daguerreotype." *Daguerreian Annual*: 1995, 148–71.

"Maison Niepce." 2020. https://photo-museum.org/the-pyreolophore-a-new-engine-principle/.

Norman, Daniel. 1938. "The Development of Astronomical Photography." *Osiris* 5: 560–94.

Stork, D. 2004. "Optics and Realism IN RENAISSANCE ART", *Scientific American*, 291(6), 76–83.

Tenn, J. S. 1985. "The Hugginses, the Drapers, and the Rise of Astrophysics." *Griffith Observer*, October Issue.

Zhang, M. 2011. "The First Portrait Ever Made." July. https://petapixel.comh/2011/07/19/the-first-self-portrait-photo-ever-made/.

CHAPTER 3

The Birth of Observational Astrophysics

PROLOGUE

By the beginning of the twentieth century, a number of the tools needed to explore space beyond the confines of the Earth had been invented and were in use by astronomers. In terms of instrumentation, these included the telescope, photography, and spectroscopy. Due to their large collecting area, telescopes could provide a dramatic improvement in light-gathering power and angular resolution over what is afforded by the human eye. Photography provided a means of both faithfully recording telescopic observations and through lengthening exposure times beyond the 1/15 second endemic to the eye, the potential of increased sensitivity. Spectroscopy provided a means of remotely determining the composition, temperature, density, and velocity of distant objects. To aid in interpreting these observations, researchers had access to Newton's laws of motion, the laws of thermodynamics, and the theory of electromagnetism. The ladder two of which were largely developed by nineteenth-century physicists, including Ludwig Boltzmann, Willard Gibbs, William Thompson, and James Maxwell. With these instrumental and theoretical tools in hand, the stage was set for cosmic discovery.

3.1 INTRODUCTION

Cave paintings and petroglyphs tell us that from our genesis humans have looked to the stars for answers about our origin and destiny. Interpretations of the night sky were often cloaked in myth and mystery. By the latter half of the nineteenth-century advances in the physical sciences (e.g., astronomy, chemistry, and thermodynamics) and technology (e.g., telescopes, photography, and spectroscopy) allowed humans, for the first time, to receive and interpret faint messages from the stars that would, in part, provide answers to the questions they had been searching for.

DOI: 10.1201/9781315210643-3

3.2 STELLAR CLASSIFICATION

As we learned in Chapter 2, the first photograph of absorption lines in the Sun's spectrum dates back to John Draper's pioneering work in 1843. As the size of telescopes and the sensitivity of photographic processes improved, it became possible to record the absorption line spectra of an ever-increasing number of stars. John Draper's son Henry and his wife Anna carried on this work at their home observatory in the United States, while William and Margaret Huggins performed spectroscopic observations of stars from their home observatory outside London.

Shortly after John Draper took the first spectrograph of the solar spectrum, Father Angelo Secchi (see Figure 3.1) was introduced to astronomical research during visiting appointments at Stonyhurst College in England and Georgetown University in the United States. After becoming director of the Roman College Observatory in 1850, he began cataloging the spectra of ~4,000 stars using an objective prism placed over the front of his telescope. If the light from stars passing through the prism was strong enough, the prism would break the light from each star into a spectrum (like small rainbows) which could be simultaneously observed, greatly increasing the number of spectra that could be recorded in a single night (see Figure 3.2). Secchi conducted his survey before photographic plates were sensitive enough to record stellar specta. So instead he drew them by hand while at the telescope. From his large sample of stars, Secchi could see similarities and differences between stellar spectra that led him to divide stars into 3 (and later 5) distinct classes (Pohle 1913), which ranged from very hot stars to cooler stars.

FIGURE 3.1 Father Angelo Secchi (1818–1878). Developed early stellar spectral classification system. Image: Public Domain.

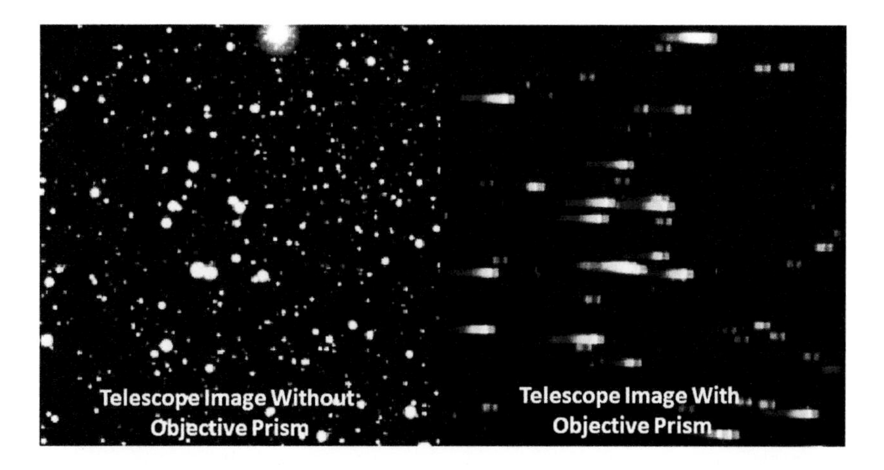

FIGURE 3.2 Spectroscopy with objective prism. Image of the Hyades star cluster without and with an objective prism over the front of the telescope. Image: Adapted from Heiter 2014.

In the 1880s, funded by an endowment from the Draper family, Edward Pickering of the Harvard College Observatory led an effort to collect and classify the spectra of as many stars as possible, with the goal of using the survey to gain insights into their properties and evolution. For this work, he used telescopes with an objective prism that fit over the telescope's aperture. This yielded spectra for every star of sufficient brightness within the telescope's field of view. The images were recorded on the most sensitive photographic plates available at the time. Pickering hired a gifted team of women to analyze and sort the spectra of the many thousands of stars that were observed. They were referred to as the "Harvard Computers" (see Figure 3.3).

Their work often included measuring the brightness, position, and color of the stars (Geiling 2017). The first installment of this work was the Draper Catalogue of Stellar Spectra, published in 1890 (Pickering 1890). Williamina Fleming (one of the computers and Pickering's former housekeeper) classified ~10,000 stars based on the strength of the hydrogen lines present in their spectra and gave them letter designations A through Q, reminiscent of Fraunhofer's designation for the hydrogen absorption lines in the Sun's spectrum (see Section 2.3). In 1897 Antonia Maury (niece of Henry Draper) and Pickering published a more detailed study of stars with bright lines in the northern hemisphere that resulted in type "B" stars being put before type "A" stars in the sequence.

By 1901 Annie Jump Cannon (see Figure 3.4) had identified regions of overlap between the spectral types, resulting in some letter designations being dropped. She also found that there was a better continuity of spectral features if type "O" stars were placed before type "B" stars. This led to the spectral type sequence O B A F G K M, which is still in use today. Each spectral class is subdivided into 10 subclasses numbered 0–9 (see Figure 3.5). The Harvard spectral classification tracks the surface (or photospheric) temperature of stars, ranging from ~100,000 K for an O star to ~2,000 K for an M star. This behavior was not fully understood until after the classification system was formulated. By 1924 over 225,300

FIGURE 3.3 Edward Pickering (1846–1919) and the Harvard Computers in front of the Harvard College Observatory (May 13, 1913). Image: Public Domain.

FIGURE 3.4 Annie Jump Cannon (left) and Henrietta Swan Leavitt (right), ca. 1913. Image: Public Domain.

FIGURE 3.5 Stellar spectral classification system. Stars are classified by the appearance of their absorption line spectrum as either O, B, A, F, K, or M, with subclasses from 0 to 9. The above image is a vertical stack of absorption line spectra from 15 different stars representing different spectral classes. The spectral classes are indicated on the vertical axis. The horizontal axis shows the relative placement of the absorption lines as a function of wavelength, with shorter (i.e., bluer) wavelengths on the left and longer (i.e., redder wavelengths) on the right. The dark vertical streaks indicate at which wavelength and to what degree atoms (e.g., hydrogen and helium) and/or molecules (e.g., carbon monoxide and vanadium oxide) of a particular element in the star's atmosphere are absorbing the continuous spectrum from the star. From the location and width of these absorption features it is possible to learn about the composition and temperature of the star. Indeed, each absorption spectrum can be thought of as a "bar code" that tells us the star's spectral type. Most stars are so distant that there is not enough light from them to trigger the color receptors (i.e., the cones) in our retinas and are, instead, detected as "white" by the more sensitive black and white receptors in the retina (i.e., the rods). "O0" stars are the hottest with surface temperatures of ~100,000 K, while "M9" stars are the coolest, with surface temperatures of ~2,000 K. With the light-gathering power of the telescope, coupled with the longer exposure time of cameras, it is possible to see the true color of stars, with the hottest stars appearing blue and the coolest red, just like the colors within a flame. Examples of stars for which we can see color with the unaided eye include Rigel, a B8 star, and Betelgeuse, an M2 star, both part of the Orion constellation. Image: NOAO/AURA/NSF.

stars had been spectrally classified by Cannon and her colleagues (Cannon and Pickering 1918–1924).

3.3 COSMIC RULERS

With a sufficiently large telescope and sensitive camera, spectroscopy could now provide a means by which the temperature and composition of stars and nebula could be determined without leaving the surface of the Earth. However, in order to begin to understand the true nature of stars, and of the universe itself, required a means of estimating distances over many lightyears. At the beginning of the twentieth century, the most advanced technique for determining astronomical distances was stellar parallax.

Parallax is the apparent shift of an object against a more distant background when observed from different vantage points. The closer the object is to the location of the vantage

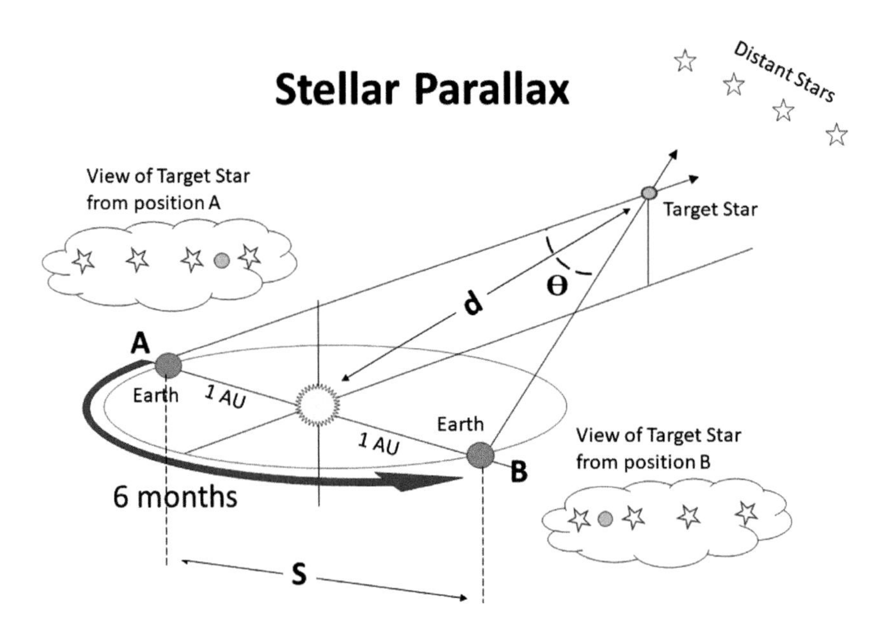

FIGURE 3.6 Stellar parallax. Our vantage point for observing the universe changes as the Earth orbits the Sun, with the result that closer objects will appear to shift their position in the sky relative to more distant objects. The shift is greatest when a target object is observed in time intervals of six months, when the Earth is on opposite sides of the Sun, corresponding to points A and B in the above figure. The separation, S, between A and B is then two astronomical units (AU). The closer an object is to the solar system, the greater the angular offset, θ, will appear relative to much more distant stars which appear fixed. Knowing S and θ, the equation of parallax can be used to calculate the distance, d, to the target object.

points, the greater the apparent shift appears to be. An example of this would be looking at your thumb at arm's length with just one eye and then the other. The shift increases as you bring your thumb closer. An analogous situation in astronomy is to observe a star against the night sky and then six months later when the Earth is on the opposite side of its orbit (see Figure 3.6). If an angular shift, θ (measured in radians) is observed relative to background stars, then knowing the diameter of the Earth's orbit, S, the distance to the star, d, can be found using the small angle formula,

$$S \approx \frac{1}{2}\theta d. \tag{3.1}$$

Aristarchus of Samos (ca. 310–230 BC) was an early advocate of the heliocentric model of the solar system and tried to use stellar parallax to determine the distances to the stars but could not detect any noticeable shift in stellar position. From this observation, he correctly concluded the stars must be very, very far away. Others, such as Tycho Brahe (~1,300 years later), were uncomfortable with the idea that such large voids actually existed and used the lack of observable stellar parallax to argue against the heliocentric model.

The diameter of the Earth's orbit is by definition two astronomical units (AU), i.e., ~3 $\times 10^{13}$ cm. If a parallax angle of 1 arcsecond is observed, substitution into Eq. (3.1) yields a distance of $d = 3.086 \times 10^{18}$ cm (or 3.26 ly), which is defined to be 1 parsec (pc). Given that the limit to the ability of the human eye to resolve objects is ~1 arcminute (60 times lower) and the nearest star is now known to be 4.24 ly away (Proxima Centauri), it is no wonder the naked eye astronomers of the past were not able to observe stellar parallax. Indeed, it was not until 1838 AD that Friedrich Bessel published the first accurate stellar parallax measurement. He used a telescopic heliometer to measure a stellar parallax of ~0.36 arcsec for 61 Cygni, which from Eq. (3.1) corresponds to a distance of 10.4 ly (Bessel 1838). The distances to Vega and Alpha Centauri were determined via parallax by Friedrich Georg Wilhelm von Struve and Thomas Henderson that same year (Hughes 2012). Even with modern instrumentation, parallax angles smaller than 0.01 arcseconds are extremely difficult to measure. This places the limit on determining distances from Earth with stellar parallax to just a few hundred light years.

The much-needed breakthrough in astronomical distance measurements was made in 1912 by one of the "Harvard Computers", Henrietta Sarah Leavitt (see Figure 3.7, *right*).

Leavitt was first hired by Edward Pickering in 1898 to measure and catalog the brightness of stars as seen in the photographic plates made by staff astronomers at the Harvard College Observatory. She briefly left the observatory to pursue her interest in art in Europe and at Beloit College in Wisconsin before rejoining the observatory in 1903, when Pickering

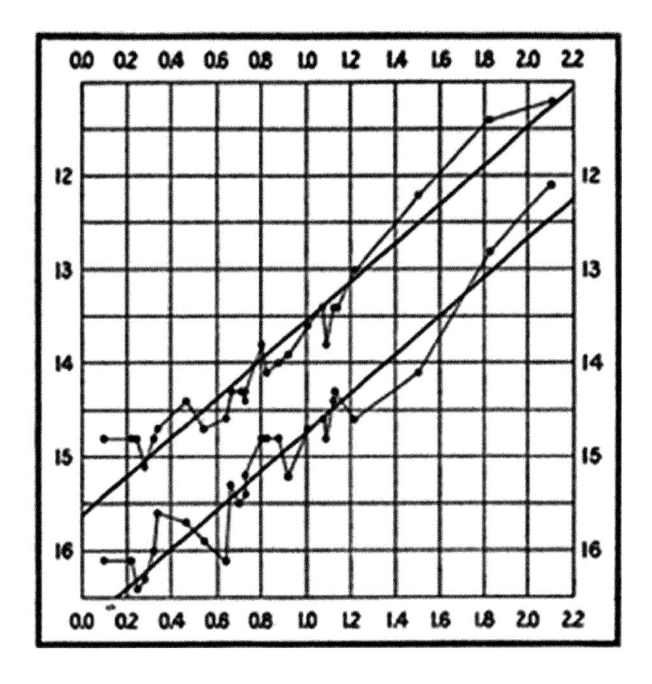

FIGURE 3.7 Cepheid variable period–luminosity relationship (from Leavitt 1912). The ordinate (vertical axis) indicates the observed maximum and minimum magnitudes of 25 Cepheid variable stars in the Small Magellanic Cloud. The abscissa (horizontal axis) is the log of the Cepheid variable's period in days. Once calibrated by observing nearby Cepheids of known distance, the observed linear relationship among maximum and minimum values could be used to determine the absolute magnitude and, subsequently, the distances to Cepheids that are hundreds, thousands, and millions of light years away.

assigned her to investigate the characteristics of variable stars on photographic plates taken in the southern hemisphere of the Small and Large Magellanic Clouds. Variable stars are stars that regularly grow brighter and dimmer over time. The time interval between when they are brightest (or dimmest) is referred to as the variable star's period. The Magellanic Clouds are dwarf galaxies that we now know are at distances of 190,000 and 160,000 light years, respectively, from the Earth. In her initial analysis of 1,777 variable stars Leavitt realized that the brightest stars had longer periods (Leavitt 1908). Sensing the potential ramifications of this result, she performed a more thorough analysis of this behavior toward a sample of 25 Cepheid variables in the Small Magellanic Cloud. She made a plot of the maximum and minimum brightness (i.e., magnitude) of each star versus its period and found that a straight line could be drawn among each of the two series of points, thereby leading her to conclude "that there is a simple relationship between the brightness of the Cepheid variables and their periods" (Leavitt 1912). The plot from her discovery paper is shown in Figure 3.7. Given that all the stars in her sample are at approximately the same distance from the Earth, she correctly concluded the relationship could, in principle, be used to deduce the *intrinsic brightness* or absolute magnitude of a Cepheid star independent of its distance. The importance of this result cannot be overstated because if one knows both the absolute, M, and apparent, m, magnitude of a star, the distance, d, to it can be found using the well-known distance modulus expression,

$$M - m = 5\log_{10}(d) - 5 \tag{3.2}$$

The expression is based on the fact that the observed brightness of an object is geometrically diluted with distance by the inverse square law. Before Leavitt's period–luminosity relation could be used together with Eq. (3.2) to determine distance, it needed to be zero-point calibrated, in effect finding the y-intercept to Figure 3.7 in terms of M, instead of m as shown. To do this required someone to determine the distances to one or more nearby Cepheid variables using stellar parallax. Within a year, Ejnar Hertzsprung did just that toward 13 Cepheids (Hertzsprung 1913). With a zero-point calibration of the Cepheid-period–luminosity relation, it was then possible to break through the confines of our solar neighborhood and begin to truly understand the size scale and nature of the universe.

3.4 MEASURING THE MILKY WAY

On a dark, clear night the Milky Way appears as a faint band of light arcing across the sky (see Figure 1.1). Its true nature has been the subject of much debate throughout human history. The ancient Greek philosopher Democritus (~400 BC), who postulated the existence of atoms, believed the universe was vast and limitless, containing innumerable worlds. He also believed the Milky Way was distant and composed of multitudes of stars: tiny, distant, compressed together, and shining brightly (Harris 2012). These are familiar concepts to us today, but they were not widely held at the time. Indeed, his contemporary, Aristotle, believed the Milky Way was smoke from stars burning in the outermost atmosphere of the Earth. In Aristotle's defense, he did believe the Earth was round, while Democritus held to the belief that it is flat. The nature of the Milky Way continued to be debated by Persian,

Arabic, and Western philosophers until 1610 when Galileo observed it with his telescope and found it to be composed of faint stars (Galileo 1610; Harris 2012). The true shape of the Milky Way was first hypothesized by Thomas Wright, an English astronomer, in 1750 when he wrote that the Milky Way's appearance was "an optical effect due to our immersion in what locally approximates to a flat layer of stars" (Wright 1750). He also believed many of the faint nebulae observable with telescopes were actually distant galaxies. These ideas were picked-up and publicized by his better-known contemporary, the German philosopher Immanuel Kant, who referred to such distant galaxies as "island universes" (Kant 1755).

In 1918, while working as an astronomer at Mt. Wilson Observatory, Harlow Shapley (see Figure 3.8) published a series of papers in which he used the new Cepheid-period–luminosity relation to determine the distances to globular clusters in the hope of determining the size and shape of the Milky Way. Globular clusters are densely packed, spherical collections of old stars that are gravitationally bound to the Milky Way. Not all globular clusters contain Cepheids. He estimated the distance to globular clusters without Cepheids by either assuming the brightest stars in those clusters or the cluster's size was the same as those in the clusters for which he could determine absolute magnitudes and sizes using the period–luminosity relation. By using "every trick in the book" he was able to estimate and plot the location of 69 globular clusters in three-dimensional space relative to the Sun.

FIGURE 3.8 Harlow Shapley (1885–1972). Successfully used the Cepheid variable period–luminosity relation to estimate the size of the Milky Way and our location within it. © AAS. Reproduced with permission. Shapley, H., 1918, "Studies Based on the Colors and Magnitudes in Stellar Clusters", Contributions from the Mt. WilsonSolar Observatory, No. 152, Preprinted from the Ap. J., Vol. XLVIII.

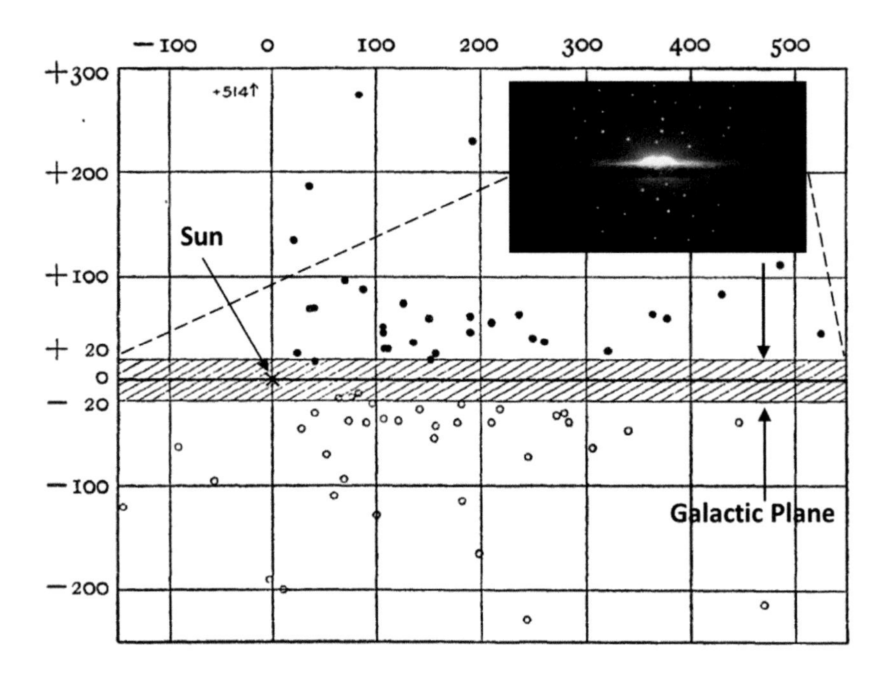

FIGURE 3.9 Harlow Shapley's original projection of the positions of 69 globular clusters on a plane perpendicular to the Galaxy. Globular clusters are shown as filled circle above the galactic plane and open circles in the southern hemisphere. The location of the Sun is at the origin (0,0), marked by an "X". The ordinate indicates distances from the galactic plane. The abscissa shows projected distances away from the Sun along the midplane. The width of the galactic plane is indicated by cross hatching. Each unit corresponds to 100 parsec. For example, a value of 500 corresponds to 500 × 100 pc = 50,000 pc. Since 1 pc = 3.26 ly, this translates to 163,000 ly. This plot informs us that (1) the Milky Way is ~200,000 ly across and (based on where the globular clusters are congregating, approximately (150,0), the Sun is far (~50,000 ly) from the galactic center. (Shapley 1918)). The inset shows a modern, artist conception of Shapley's Milky Way schematic (adapted from Young 2020).

The resulting figure from his 1918 paper is shown in Figure 3.9. From this modest-looking plot, Shapley estimated the diameter of the Milky Way to be an astonishing ~230,000 light years. This value is within a factor of two of the modern estimate. His larger number was due to him not taking into account the dimming effects of interstellar dust which was not understood at the time. In interpreting the figure Shapley correctly assumed that most of the globular clusters would be found toward the center of the galaxy, where the gravitational pull would be expected to be greatest. From his plot he found that our Sun is not located in the center of the Milky Way but ~50,000 light years out in the galactic boondocks. This result further strengthened the Copernican view of cosmology, in which the Earth, and now the Sun, is not at the center of anything. Our star system is just one of ~400 billion others strewn throughout the Milky Way.

3.5 MEASURING THE UNIVERSE

In 1917, while Shapley was working to publish his pivotal papers on the size of the Milky Way, his colleagues at the Carnegie Institute Mt. Wilson Observatory were commissioning

FIGURE 3.10 Edwin Hubble's (1889–1952) WWI Military ID. This picture was taken in 1918, just a year before he joined the staff at the Mt. Wilson Observatory.

a new 100-inch telescope, the most powerful in the world. With its unparalleled light-gathering power, it would be possible to use the Cepheid-period–luminosity relation to measure distances on cosmic size scales. However, in 1921 Shapley left Mt. Wilson to replace the recently deceased Edward Pickering as the director of Harvard College Observatory.

The task of utilizing the 100-inch telescope for distance determination fell to a former lawyer, high-school teacher, and athlete, Edwin Hubble, who had received a PhD in astronomy from the University of Chicago in 1917. After this he joined the US Army to participate in World War I, but the war ended before he saw action (see Figure 3.10). Upon being discharged in 1919, he was offered an astronomer staff position at Mt. Wilson by the observatory director, George Hale. Upon arriving at Mt. Wilson, Hubble met Milton Humason (see Figure 3.11).

Humason, like Hubble, had an unconventional background. He had no formal education beyond the age of 14. He left home in 1905 to work at the Mount Wilson Hotel. He also worked as a "mule skinner", hauling building materials up to what was to be the Mt. Wilson Observatory (see Figure 3.12). While working on Mt. Wilson, he met and in 1911 married the daughter of the observatory's chief electrician. Soon after, he took a foreman job on a relative's ranch on the valley floor. Six years later, he returned to the observatory and took the job of *janitor*. Out of curiosity, he volunteered as a telescope night assistant, aiding staff astronomers with their observations. He performed so well in this role, that in 1919 George Hale made him an observatory staff member just in time to begin to work with Hubble (Simmons 2021).

3.5.1 The Trouble with Andromeda

On a dark, clear night in the northern hemisphere, the Great Nebula of Andromeda (also known as M31) can be observed with the unaided eye. It was described as a "nebulous

FIGURE 3.11 Milton Humason (1891–1972). He was a gifted observer and Mt. Wilson staff member who teamed with Hubble.

FIGURE 3.12 A 60-inch telescope being pulled up the Mt. Wilson Rd. by a team of mules, *ca.* 1908. Humason (believed to be standing, far left) started out as a mule driver or "skinner".

smear" by the Persian astronomer Abd al-Rahman al-Sufi around AD 964 (Hafez 2010). Telescopic observations of Andromeda were first reported by Simon Marius (a contemporary rival of Galileo) in 1612, who described its appearance as a dull pale light which increased in brightness toward its center (Bond 1848). By the mid-eighteenth century, the idea that Andromeda is an "island universe" composed of a collection of stars was being discussed by European astronomers, such as Pierre Louis Moreau de Maupertuis, Thomas Wright, and Immanuel Kant (Kant 1755). Before the 100 inch Hooker telescope, the largest telescope in the world was the Leviathan telescope built in Ireland by William Parsons, the third Earl of Rosse. The Leviathan had a 6 ft clear aperture and was 53 ft long. Using the Leviathan, Parsons was able to make out the spiral nature of Andromeda and other similar nebulae (King 1855; Parsons 1850). In 1864, using a spectroscope attached to a telescope, Sir William Huggins (see Chapter 2) observed the spectrum of Andromeda. He found that it was not composed of bright emission lines like other nebulae, but had in fact an absorption line spectrum similar to the Sun's, supporting the idea that Andromeda is a collection of stars (Huggins 1864). The first photograph to clearly show the spiral structure of Andromeda was taken in 1888 by Isaac Roberts (see Figure 3.13).

By 1912 spectrographs and cameras had improved in performance to the point that Vesto Slipher (see Figure 3.14) was able to measure the Doppler shift in the spectral lines of nebulae (like Huggins had done ~50 years before toward the star Vega, see Chapter 2). For Andromeda he found the Doppler shift indicated it was moving toward us at a record-setting velocity of 300 km/s, further indicating that it was no normal nebula. When Slipher made the measurement there was still no known way to accurately measure the distance to Andromeda. So, the jury was out on whether objects like Andromeda were within the Milky Way or external galaxies, the "island universes" of Wright and Kant.

FIGURE 3.13 Great Nebulae in Andromeda. This long exposure photograph of Andromeda was taken by Isaac Roberts in 1888 with a 20-inch diameter (100-inch long) reflecting telescope. This was the first photograph to clearly show the spiral structure of Andromeda. (A later version would become a very popular computer screen saver.)

FIGURE 3.14 Vesto Slipher (1875–1969). Using spectroscopy, he was the first to catalog the velocities of "spiral nebulae", which were soon determined to be distant galaxies by Hubble and Humason. Image from a 1905 family photo.

The truth was coming. In 1917 Heber Curtis observed a nova in Andromeda. A nova is a bright star that appears suddenly and then fades away over several weeks or months (see Chapter 4). At that time the true nature of novae was not known, but he noted that the one in Andromeda was about ten magnitudes dimmer than all others that had been seen. He correctly concluded that Andromeda must be very far away. He estimated its distance to be ~500,000 ly, which supported the island universe theory, namely that spiral nebulae like Andromeda were actually distant galaxies in their own right, independent of the Milky Way (Curtis 1917). Curtis argued for this interpretation to good effect in the "Great Debate" on the nature of spiral nebulae held in 1920. His opposition in the debate was none other than Harlow Shapley, who believed the spiral nebulae were on the outskirts of our own galaxy. In 1922, Ernst Opik calculated the distance to Andromeda by estimating its mass and luminosity based on the rotational velocity of stars about its center. This put it at 450,000 pc, or ~1.5 million light years (Opik 1922). Although this estimate is not far off from the modern value of 2.5 million light years, the assumptions he employed in his analysis were not widely accepted. Just a few years later Edwin Hubble used photographic plates of Andromeda taken with the Mt. Wilson 100-inch and 60-inch telescopes to identify and measure the periods of a dozen Cepheid variables within Andromeda. He then used the Cepheid variable period–luminosity relation, originally found by Leavitt and later improved upon by Shapley, to determine a distance to Andromeda of 285,000 parsec, just under 1 million light years (Hubble 1925). Although it would later be found that Opik's estimate is closer to reality, Hubble's distance estimate based on the Cepheid-period–luminosity relation was widely accepted and ended the debate on whether external galaxies existed and set the stage for the revelations to come.

3.5.2 Observational Support of the Big Bang

After measuring the velocity of Andromeda in 1912, Vesto Slipher continue to use the spectrograph on the 24-inch refracting telescope at the Flagstaff Observatory (together with a new, more sensitive camera) to measure the relative velocities of both "bright-line" nebulae (like the Orion Nebula) and "dark-line" nebulae (like Andromeda). Slipher described dark-line nebulae as being "faint in the extreme." After two years of painstaking efforts, he published velocities toward 15 spiral nebulae. He found that 12 of the 15 spiral nebulae had positive (i.e., redshifted) velocities, indicating they were moving away from us (Slipher 1915). Two of them had redshifted velocities of + 1100 km/s. In 1915 the distances to these objects were still unknown. However, by 1929 their extragalactic nature had been firmly established and radial velocities toward 46 of them measured. By estimating the absolute magnitudes of Cepheid variables, novae, and the brightest stars in these galaxies, Hubble and Humason were able to estimate distances toward 22 of them. They then made an x-y plot with each galaxy's velocity being indicated on the y-axis and its distance on the x-axis. The resulting plot from Hubble's 1929 paper is shown in Figure 3.14. Even though the data is, as Hubble said, "scanty … and poorly distributed", the plot clearly indicates a general linear dependence of a galaxy's velocity (v) with its distance (d) such that

$$v = H_0 d \tag{3.3}$$

where H_0 is the slope of the line and referred to as Hubble's constant. In his paper Hubble did not suggest that the observed linear dependence was the result of the universe expanding, but rather it may be the result of time slowing down at great distances, known as the de Sitter effect, a consequence of the de Sitter static universe model, which has long since been abandoned (Bachall 2015).

The expansion of the universe was first proposed in 1922 by the Russian cosmologist Alexander Friedmann (see Figures 3.15 and 3.16) as a natural consequence of Einstein's Theory of General Relativity (Einstein 1915). Einstein had remarked on this possibility in a 1917 paper, but since at that time there was no observational evidence for such an expansion, he deemed it a non-physical solution to his equations. Instead, he proposed the concept of a cosmological constant (Λ) to counter balance the effects of gravity and hold the universe in a steady state (Einstein 1917). Friedmann discussed the consequences of an expanding universe in a series of papers in 1924 (Friedmann 1924). His solutions are referred to as the Friedmann equations and are a cornerstone of modern cosmology. Unfortunately, Friedman died a year later from typhoid fever, which he is believed to have contracted by eating an unwashed pear he bought at a train station on his way back from a vacation in Crimea (Falkovich 2011). Working independently, Georges Lemaitre (see Figure 3.17), a Belgian Roman Catholic priest who earned his

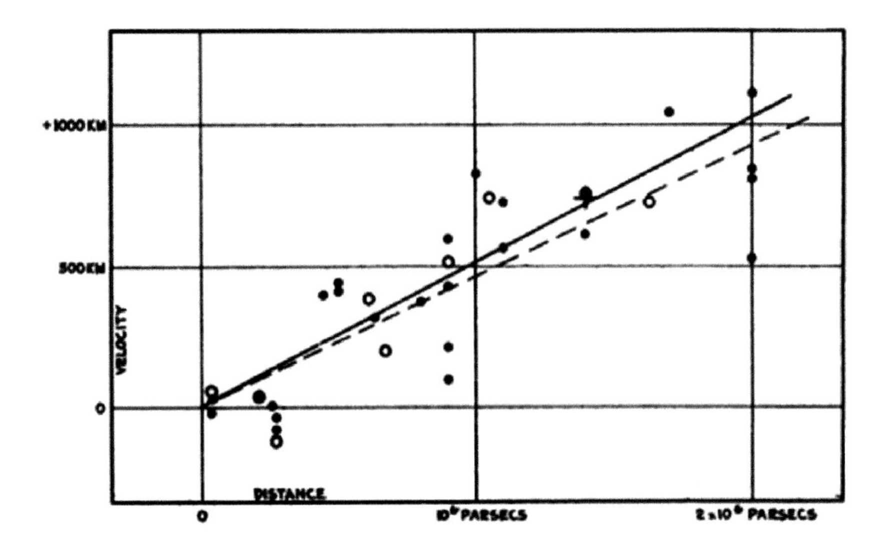

FIGURE 3.15 "Velocity-Distance Relationship for Extragalactic Nebulae". In the above figure the observed velocity of galaxies is plotted as a function of their distance. Even though the data is sparse, it clearly shows that the more distant a galaxy, the faster it is moving away from us. When published in 1929, it was the strongest observational evidence to date for the expansion of the universe. Image credit: Hubble, E, 1929, "A relation between distance and radial velocity among extra-galactic nebulae". PNAS. 15 (3): 168–173.

PhD in physics at MIT, reached a similar conclusion as Friedmann and published his results in 1927 (Lemaître 1927). Lemaitre was the first to propose that an expanding universe could explain the redshift of the galaxies observed by Slipher, although at that time his results were inconclusive. He published his results two years before Hubble. Lemaitre communicated his results to Einstein, who commented, "Your calculations are correct, but your physics is atrocious" (Deprit 1984). Einstein, who was a firm believer in the steady-state model of the universe, found Lemaitre's observational evidence unconvincing. Einstein's reversed his opinion in 1929 with the publishing of Hubble and Humason's observations (Figure 3.15). In 1931 Lemaitre took the idea of an expanding universe a step further and proposed that if you ran the expansion backward in time, it would shrink to a single point. It is at this single point in time that the universe had its beginning. An estimate for when this occurred, i.e., the age of the universe, can be obtained by simply taking the reciprocal of Hubble's constant, H_0,

$$\text{Age of Universe} = \frac{1}{H_0} \tag{3.4}$$

This concept for the origin and evolution of the universe is now known as the Big Bang model. The term "Big Bang" was not coined by Friedmann, Lemaitre, or Hubble but by the British astronomer Fred Hoyle during a 1949 BBC radio broadcast (Hoyle 1949). Ironically, Hoyle was a critic of the theory and remained so until his death in 2001. In recognition of Lemaitre's pioneering work, in 2018 the International Astronomical Union voted to add Lemaire's name to Hubble's and refer to the observed relationship between the velocity and distance of galaxies as the "Hubble-Lemaitre Law" (IAU 2018).

FIGURE 3.16 Alexander Friedmann (1888–1925) was the first to promote a solution to Einstein's General Theory of Relativity that suggested the universe is expanding. He was a pilot in the Russian Imperial Army during WWI and set the altitude record of 24,420 ft in a balloon in 1925. He died before his model was observationally confirmed by Hubble (Frenkel 1988).

In the century since Slipher, Hubble, and Humason published their initial results based on just 24 nearby galaxies, 1000s of much higher-quality measurements of the velocities and distances of galaxies have been made, many of these with the Hubble Space Telescope (HST). Figure 3.18 is a more modern version of Hubble's original plot based on observations of Type Ia Supernovae at distances 100's of time greater than those observed by Hubble himself. The 24 galaxies he and Humason observed would fit in a box near the origin of the plot. The linear relationship Hubble observed still holds, even though due to calibration errors his original distance estimates were low by a factor of ~7. His value of H_0 was 500 km/s/Mpc. This is in contrast to the modern estimate of $H_0 = 70 \pm 2$ km/sec/Mpc. Using the above expression, the age of the universe can be estimated to be 13.8±0.1 billion years (Bachall 2015). However, the story does not end here.

After Hubble's observations showing the universe is expanding, Einstein dropped his concept of a cosmological constant, in effect, setting its value to zero. This was accepted by the world's cosmologists until 1998, when observations made with HST of supernovae explosions in galaxies at high redshift indicated that the expansion of the universe is accelerating at great distances (Schaefer et al. 1998; Schmidt et al. 1998). Such a behavior

FIGURE 3.17 Georges Lemaitre (center), Robert Millikan (left), and Albert Einstein (right). This picture was taken in 1933 just after Lemaitre had given a talk on his theory for the origin of the universe at the California Institute of Technology. Einstein had first rejected the expansion theories of both Friedmann and Lemaitre, instead endorsing the concept of a steady-state universe. Once Hubble's observations supporting an expanding universe were published, Einstein quickly endorsed the work of Lemaitre and Friedmann. Indeed, after Lemaitre's talk at Caltech, Einstein stood up and publicly praised his work (Kragh 1999). Millikan (an experimental physicist) had earlier measured the charge of the electron and estimated Planck's constant. At the time of the photograph, he was president of Caltech.

could be explained if our universe has a positive cosmological constant, Λ. This would indicate that our universe is permeated with a low-level of vacuum energy, also called dark energy. Estimates of the density of this energy can be made from the fact it is only observable at great distances, where normal matter has thinned-out enough for dark energy to have an effect. At these distances dark energy causes the above plot to "bend-up" at the far right. Detailed fits to the data indicate our universe consists of 68.3% dark energy, 26.8% dark matter (which has gravity, but does not interact with light), and 4.9 % normal matter (which includes us). These results indicate the universe will continue to expand and cool forever. In about 2 trillion years all but the closest galaxies will be beyond the observational horizon. About 100 trillion years after that the last stars will go out and the universe will slip into eternal darkness. Long before this, the survival instinct inherent to whatever intelligence species remains will have motivated them to seek refuge elsewhere.

SUMMARY

The historical arc of the twentieth century is often described in terms of the geopolitical conflicts that consumed much of mankind's energy and resources during this period. It was also a time of great scientific and technological advances. Some of these advances were motivated by the conflicts themselves. At the start of the century, we had only a dim awareness of the size and nature of the universe. Many of these ideas had been tossed around, at least qualitatively, since the time of the ancient Greeks. It was not until the twentieth

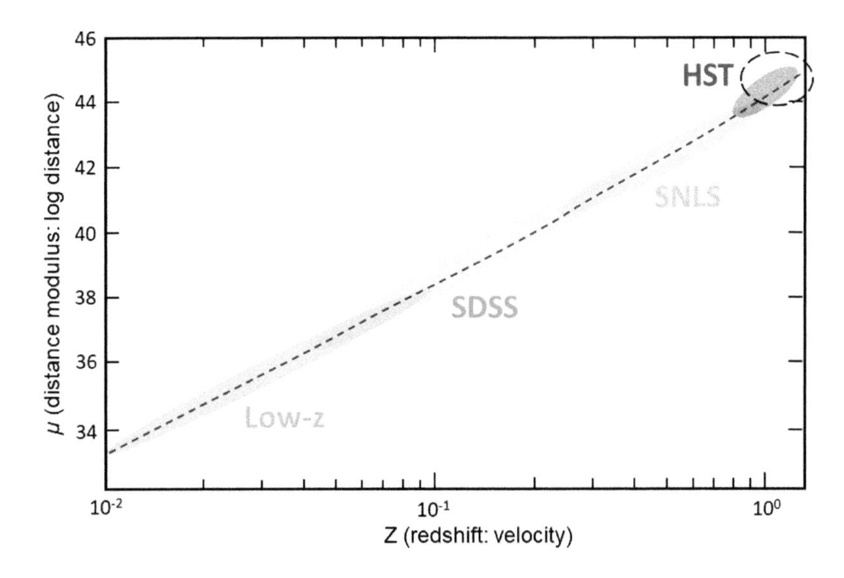

FIGURE 3.18 Modern Hubble diagram made from observations of Type I Supernovae in galaxies. (The axes are shown reversed compared to Figure 3.14.) The plot confirms the trend observed by Hubble and Humason, namely that the recessional velocity of galaxies continues to increase with distance. Different colored ellipses indicate different samples of galaxies. (Low-z sample; Sloan Digital Sky Survey Sample (SDSS); Supernova Legacy Survey (SNLS); Hubble Space Telescope Survey (HST)). The quality of the fit to the dashed black line is consistent with a cosmology where the universe is flat with a matter density (dark + ordinary) of ~30% and a cosmological constant representing dark energy of ~70%. The upward deviation from the line at high redshift (e.g., see dashed ellipse) is evidence for galaxy acceleration. Hubble's original plot (Figure 3.15) would fit in a small dot near the origin of this diagram. Data from Bahcall 2015.

century that there was the confluence of theory and technology required to empirically test, mature, and (in many instances) discard old ideas in favor of new ones. The two key twentieth-century breakthroughs in theory were Einstein's Theory of Relativity (Special and General) and quantum mechanics (which Einstein was not a fan of). More powerful telescopes, combined with advances in photographic and spectroscopic techniques, allowed observers in the first half of the century to quickly determine that our home here on Earth is very, very small in the scheme of things. Further advances in telescope and detector technology continued to push the observational limit, culminating in a small fleet of space telescopes that pushed the observable horizon to the very edge of the universe. Each time a new generation of detector or telescope went into operation, a new window to the universe was opened, often leading to unexpected discoveries. We will take a peek at more of these discoveries in the coming chapters.

REVIEW QUESTIONS

1) What made the spectroscopic observation of hundreds of thousands of stars possible?

2) What is the stellar classification system based on?

3) If a star has a measured parallax of 0.5 arc seconds, how far away is it in light years? (Hint: see Appendix 1 for unit conversions).

4) Which catholic priest developed the first stellar classification system based on the observation of 4,000 stars?

5) Who was the former housekeeper of Pickering that spectrally classified 10,000 stars?

6) Who led the classification of ~225,000 stars?

7) Who discovered the Cepheid-period–luminosity relationship? Why is it so important?

8) Who first measured the Milky Way? How was it done?

9) Who first measured the velocities of galaxies?

10) Who was the former "mule skinner" and janitor that worked with Hubble?

11) Who was the former WWI fighter pilot and record-setting aeronaut that first solved Einstein's field equations and predicted the universe is expanding? What did Einstein think of his results? How did he die?

12) Which catholic priest predicted the expansion of the universe from Einstein's equations and compared his results to observations? What did Einstein initially think of his results? Did Einstein ever change his mind?

13) What are the units of the parameters within the Hubble-Lemaitre Law? Use dimensional analysis (i.e., tracking units as a calculation is performed) to show how the reciprocal of the Hubble constant is the age of the universe.

14) What does a modern Hubble diagram tell us about the ultimate fate of the universe?

15) From an analysis of the behavior of a modern Hubble Diagram at great distances, what has been deduced about the composition of the universe?

REFERENCES

Bahcall, N. 2015. "Hubble's Law and the Expanding Universe." *PNAS* 112 (11): 3173–5.

Bessell, F. W. 1838. "Bestimmung Der Entfernung Des 61sten Sterns Des Schwans." [Determination of the Distance to 61 Cygni]. *Astronomische Nachrichten (in German)* 16 (365–366): 65–96.

Bond, G. P. 1848. "An Account of the Nebula in Andromeda." *Memoirs of the American Academy of Arts and Sciences*, New Series Volume 3: 75–6.

Cannon, A., and E. Pickering. 1918. "The Henry Draper Catalogue." *Annals of Harvard College Observatory.*

Curtis, H. 1917. "Novae in Spiral Nebulae and the Island Universe Theory." *Publications of the Astronomical Society of the Pacific* 29 (171): 206–7.

Deprit, A. 1984. "Monsignor Georges Lemaître." In *The Big Bang and Georges Lemaître*, edited by A. Barger, 370. Reidel.

Einstein, A., 1915. https://einsteinpapers.press.princeton.edu/vol6-trans/110.

Einstein, A. 1917. "Kosmologische Betrachtungen zur Allgemeinen Relativitätstheorie." *Sitzungsberichte der Königlich Preußischen Akademie der Wissenschaften*. Berlin, D. E. part 1:

142–152. Bibcode:1917SPAW....... 142E. Archived from the original on 2019-03-21. Retrieved 2014-11-15.

Falkovich, G. 2011. "The Russian School." https://www.semanticscholar.org/paper/The-Russian-school-Falkovich/a819ff4616bef5312ec85da19e7c4679cccff471.

Frenkel, V. 1988. "'Aleksandr Aleksandrovich Fridman (Friedmann): A Biographical Essay'." *Soviet Physics Uspekhi* 31 (July): 645–65. https://ufn.ru/ru/articles/1988/7/d/.

Friedmann, A. 1924. "On the Possibility of a World with Constant Negative Curvature of Space." Published by the German Physics Journal Zeitschrift *für* Physik 21(1): 326–332.

Galilei, Galileo. 1610. *Sidereus Nuncius*, 15 and 16. Venice, Italy: Thomas Baglioni.

Geiling, Natasha. 2017. "The Women Who Mapped the Universe and Still Couldn't Get Any Respect." *Smithsonian.com*, September 18, 2013. Accessed October 12, 2017.

Hafez, Ihsan. 2010. *Abd al-Rahman al-Sufi and His Book of the Fixed Stars: A Journey of Re-Discovery* [Ph.D. Thesis]. James Cook University.

Harris, L. 2012. "Visions of the Milky Way in the West: The Greco-Roman and Medieval Periods." In *Culture and Cosmos*, edited by Nicholas Campion, and Rolf Sinclair 16 (1 and 2): 271–82.

Heiter, U. 2014. "Rainbows in Light." November. https://www.astro.uu.se/~ulrike/Spectroscopy/PPT/Hyaden.GIF.

Hertzsprung, E. 1913. "Über die Räumliche Verteilung der Veränderlichen vom δ Cephei-Typus" ["On the Spatial Distribution of Variable [Stars] of the δ Cephei Type."] *Astronomische Nachrichten (in German)* 196 (4692): 201–8.

Hoyle, F., 1949. https://www.joh.cam.ac.uk/library/special_collections/hoyle/exhibition/radio/.

Hubble, E. 1925. "Cepheids in Spiral Nebulae." *The Observatory* 48,p. 139–142.

Hubble, E. 1929. "A Relation between Distance and Radial Velocity among Extra-Galactic Nebulae." *PNAS* 15 (3): 168–73.

Huggins, W. 1864. "On the Spectra of Some of the Nebulae." *Philosophical Transactions of the Royal Society of London* 154: 437–44.

Hughes, Stefan. 2012. "Catchers of the Light." *ArtDeCiel Publishing*: 702.

IAU. 2018. "Press Release iau1812." https://www.iau.org/news/pressreleases/detail/iau1812/.

Kant, I. 1755. *Allgemeine Naturgeschichte und Theorie des Himmels, Part I*, J.F. Peterson, Königsberg and Leipzig.

King, H. C. 1955. *The History of the Telescope*, 206–17. High Wycombe: Charles Griffin and Co.

Kragh, H. 1999. Cosmology and Controversy: The Historical Development of Two Theories of the Universe. Princeton University Press. ISBN 0-691-00546-X.

Leavitt, H. 1908. "1777 Variables in the Magellanic Clouds." *Annals of Harvard College Observatory* 60: 87–108.

Leavitt, H., and E. Pickering. 1912. "Periods of 25 Variable Stars in the Small Magellanic Cloud." *Harvard College Observatory Circular* 173: 1–3.

Lemaître, Georges. 1927. "Un Univers Homogène de Masse Constante et de Rayon Croissant Rendant Compte de la Vitesse Radiale des Nébuleuses Extra-Galactiques." *Annales de la Société Scientifique de Bruxelles (in French)* 47: 49–59.

Opik, E. 1922. "An Estimate of the Distance of the Andromeda Nebula." Astrophysical Journal 55: 406–10.

Parsons, William (Lord Rosse). 1850. "Observations on the Nebulae." *Philosophical Transactions of the Royal Society* 140: 499–514.

Pickering, Edward C. 1890. "The Draper Catalogue of Stellar Spectra Photographed with the 8-Inch Bache Telescope as a Part of the Henry Draper Memorial." *Annals of Harvard College Observatory* 27: 1–388.

Pohle, J. 1913. "Angelo Secchi." In Catholic Encyclopedia, edited by Charles Herbermann. New York: Robert Appleton Company.

Schaefer, B., N. Walton, M. D. Valle, S. Deustua, R. S. Ellis, S. Fabbro, A. Fruchter, G. Goldhaber, D. E. Groom, I. M. Hook, A. G. Kim, M. Y. Kim, R. A. Knop, C. Lidman, R. G. McMahon,

P. Nugent, R. Pain, N. Panagia, C. R. Pennypacker, P. Ruiz-Lapuente, B. Schaefer, and N. Walton. 1998. “Discovery of a Supernova Explosion at Half the Age of the Universe.” *Nature* 391 (6662): 51–4.

Schmidt, B. P., N. B. Suntzeff, M. M. Phillips, R. A. Schommer, A. Clocchiatti, R. P. Kirshner, P. Garnavich, P. Challis, B. Leibundgut, J. Spyromilio, A. G. Riess, A. V. Filippenko, M. Hamuy, R. C. Smith, C. Hogan, C. Stubbs, A. Diercks, D. Reiss, R. Gilliland, J. Tonry, J. Maza, A. Dressler, J. Walsh, and R. Ciardullo. 1998. “The High-Z Supernova Search: Measuring Cosmic Deceleration and Global Curvature of the Universe Using Type Ia Supernovae.” *Astrophysical Journal* 507 (1): 46–63.

Simmons, M. 2021. “Building the 100 Inch Telescope.” https://www.mtwilson.edu/building-the-100-inch-telescope/.

Slipher, V. M. 1915. “Spectroscopic Observations of Nebulae.” *Popular Astronomy* 23: 21–4.

Wright, T. 1750. *An Original Theory or New Hypothesis of the Universe...* . London: H. Chapelle. From pages 48.

Young, M. 2020. “Star Clusters Reveal the Krakem in the Milky Way.” *Sky & Telescope*. Accessed November 13, 2020. https://skyandtelescope.org/astronomy-news/stellar-fossils-reveal-the-kraken-in-the-milky-ways-past/.

CHAPTER 4

Stellar Evolution

PROLOGUE

We are children of the Sun. The Earth itself is a byproduct of the Sun's creation from interstellar matter approximately 4.6 billion years ago. It is the light from the Sun that keeps much of the Earth's surface above freezing and has driven the evolution of life. It has been only ~60 years, less than an average human lifetime, since we evolved to the point where we are capable of communicating between the stars. If the origin and evolution of life on Earth is typical of what happens elsewhere, then this implies that for a star to be a "good star" it must be capable of providing a stable environment for one or more of its planetary offspring over the billions of years required for the evolution of intelligent life. If this is indeed true, then several questions immediately come to mind. How unique is the Sun among the stars? How long will the Sun live? How many Sun-like stars are there? What is the source of a star's energy? The ability to travel to the stars is likely to remain out of reach for generations to come. How then can we hope to answer these fundamental questions about their (and our) existence? Fortunately, the same observational and theoretical tools employed by astronomers at the beginning of the twentieth century to determine the origin of the universe could be used to address the mystery of stellar evolution as well.

4.1 INTRODUCTION

The true nature of stars has been a subject of speculation since prehistoric times. With no artificial lights to mask their presence, many stars that are too dim to be observed today except from remote locations were routinely available to our ancestors on any clear, moonless night. The nightly motions of celestial objects were used throughout human history to tell time and predict the comings and goings of the seasons, thereby indirectly governing the activities of their human observers. It is perhaps not surprising then that over the centuries, the seemingly eternal and predictive nature of stars led our ancestors to bestow godlike qualities on them, with the result being they often played significant roles in many religions, myths, and legends. This was particularly true of the handful of celestial objects that fell out of step with the nightly parade of stars, sometimes even appearing to move

DOI: 10.1201/9781315210643-4

backward. These were the "wanderers" (or "planete" in Greek), which appear to have the power to follow their own path. Today we know them as Mercury, Venus, Mars, Jupiter, and Saturn. A human need to find a connection to the stars appears engrained in our collective psyche. Even now, when the true nature of stars and planets is well known, astrology is used by millions of people each day as a source of amusement, direction, and/or meaning.

4.2 CATALOGING THE STARS

In science, when one wishes to understand the nature of an object or phenomenon, the first step is to perform systematic observations. This was true for stars as well. The oldest known star charts can be found on the southern ceiling panel (see Figure 4.1) of an Egyptian tomb in Thebes on the west bank of the Nile. The tomb belonged to Senmut, the royal vizier and calendar keeper of Queen Hatshepsut. The southern panel depicts the conjunction of planets in the vicinity of the star Sirius. A modern planetarium program can be used to show that such a configuration occurred over Egypt around May 1, 534 BC (von Spaeth 2000). The first known star catalog was the "Three Stars Each", compiled by Babylonian astronomers around 1,200 BC.

The catalog contains positions for 36 stars, which were used to divide the sky into three regions: northern, equatorial, and southern. The Babylonians utilized a base 60 system to divide angles and time into hours, degrees, minutes, and seconds. This system was originally developed by their Sumerian predecessors around ~3,000 BC and is still in use today. A clay tablet of Babylonian astronomical data is shown in Figure 4.2.

Arguably, the greatest astronomer of antiquity was Hipparchus (see Figure 4.3). He was born in 190 BC in the Greek city of Nicaea, in what is now northern Turkey. He was an

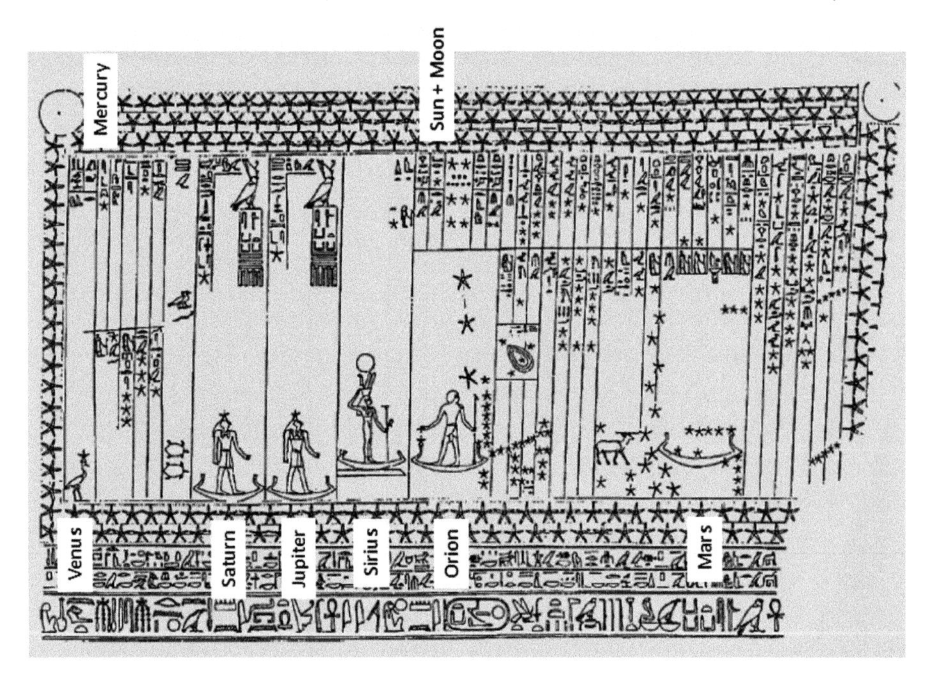

FIGURE 4.1 Star Chart from the tomb of Senmut, ca. 1,534 BC. Adapted from von Spaeth 2000; used with kind permission.

FIGURE 4.2 Tablet 1 of the *Mul-Apin* ("the plough star"). The Babylonian text divides the sky into three parts and includes the rising of principal stars and constellations in the path of the moon. Image 152339001: British Museum, used with permission.

FIGURE 4.3 Hipparchus (190–120 BC) on Roman Coin. Hipparchus is generally considered the greatest of the ancient astronomers. He is sometimes featured on the back of Roman coins, such as this one (ca. 253–260 AD) for Emperor Valerian (left). Hipparchus (right) is seated next to a globe (presumably the Earth) on a pedestal. Image courtesy of Ralph DeMarco Ancient Coins.

accomplished mathematician and is considered the father of geometry. He constructed tables of trigonometric functions and made advancements in spherical geometry, which he then used to make the first mathematical models of the motions of the Sun, Moon, and planets. (Isaac Newton was motivated to invent calculus for the same reasons.) He measured the positions of more than 850 stars. Some of the stars in his catalog had also been observed by earlier Babylonian astronomers. He found a systematic shift of ~2° (about two finger widths) in the positions between the stars in his catalog compared to those of the Babylonians (Mark 2022; Toomer 1988). He correctly deduced that this shift was due to a "wobble", or precession, of the Earth's position relative to the fixed stars. He also noted that the lengths of the seasons are not equal, a fact that could not be accounted for if the Sun was in a circular orbit moving at a constant velocity about the Earth (Jones 2021). This deduction led him to consider a heliocentric model of the solar system where it is the Sun, not the Earth, that is at the center. Although such a model was being discussed by Aristarchus of Samos and others two centuries earlier, Hipparchus was among the first to attempt the development of a mathematical framework to support it. Unfortunately, any orbit other than circular went against the belief, earlier exalted by Aristotle, that all things in the heavens were eternal and perfect, with the perfect motion being circular. The requirement for a noncircular orbit to explain his observations troubled Hipparchus and led him to abandon the heliocentric model in favor of an Earth-centered, geocentric one (Mark 2022). It was another ~1,800 years before Johannes Kepler, using observations of the motions of Mars provided to him by Tycho Brahe, unequivocally showed that planetary orbits are elliptical and the speed at which planets travel along them varies with time. Kepler's results indicated that Hipparchus' original calculations for a heliocentric model were pointing toward the truth. This story is reminiscent of that of Einstein and his Theory of General Relativity, where, if Einstein had followed his own theory without prejudice, it may have led him to predict the Big Bang years before it was observed (see Section 3.5.2).

4.3 THE COLOR OF STARS

By the early twentieth century, star catalogs contained not only the positions, but also, thanks to the work of Annie Jump Cannon and the other Harvard "computers", the spectral types of over a hundred thousand stars. As discussed in Chapter 3, based upon the appearance of their absorption line spectrum, 99% of all stars can be classified as being either O, B, A, F, G, K, or M, with subclasses from 0 to 9 (see Figure 4.4). A mnemonic device for remembering stellar spectral types is credited to American astronomer Henry Russell and is described in Figure 3.5. The absorption lines arise in the outer and least dense layers of stellar atmospheres. From the absorption line patterns it is possible to deduce the elemental composition (e.g., hydrogen and helium) of stars by employing Kirchoff's laws (see Section 2.5). On one end of the spectral classification scale are the blue O0 stars, with surface temperatures of ~100,000 K. On the other end of the scale are the red M9 stars, with surface temperatures of ~2,000 K. Between spectral types O and M, stars representing all colors of the rainbow can be observed, with the hottest stars appearing blue and the coolest stars appearing red, for the same reason very hot objects appear blue or "white hot", while cooler objects, like the dying embers of a fire, appear red. (All but the brightest stars

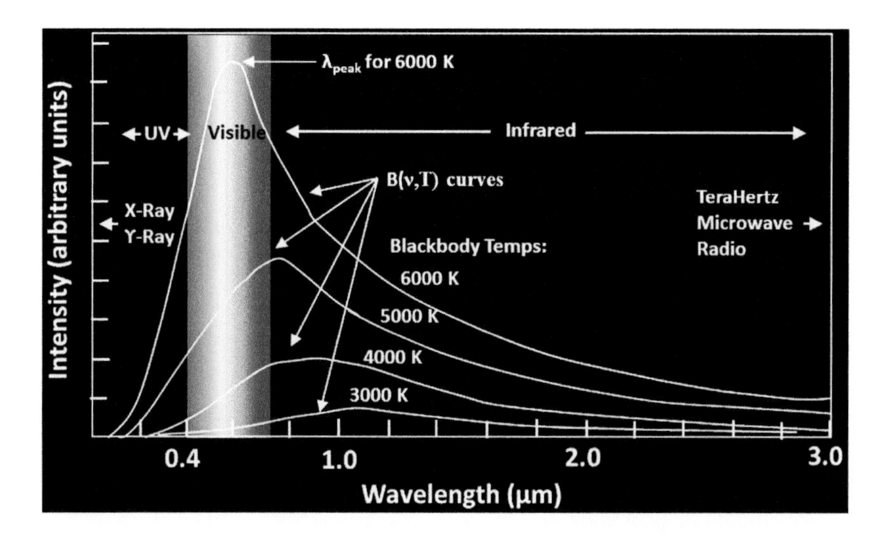

FIGURE 4.4 Plots of the Planck Function for various temperatures found on the surfaces of stars. The wavelengths and temperatures at which blackbody emission is perceived by the human eye are shown in color. Our eyes are blind to the rest. The terahertz, microwave, and radio parts of the spectrum are at longer wavelengths off the righthand side of the chart.

appear white to us because of the reduced sensitivity of the color receptors (i.e., cones) in our eyes.) The underlying reason for the relationship between the temperature and color of an object was first described by Max Planck in 1900 and led him and others to the conclusion that the light we see is not a continuous wave of energy but is instead composed of a stream of particles, i.e., photons. Planck concluded that each photon contains an amount of energy proportional to its frequency, as described by his famous equation,

$$\Delta E = h\nu = h\frac{c}{\lambda} \tag{4.1}$$

where

ΔE = the energy within a photon (joules)
ν = the frequency of a photon (Hz)
λ = wavelength (*i.e.*, characteristic size) of photon (m)
h = Planck's constant (6.626×10^{-34} joule-sec)

Photons that appear blue contain more energy and have higher characteristic frequencies than photons that appear red. The possibility that light is composed of particles or "corpuscles" had been discussed at length by Rene Descartes and Isaac Newton over two centuries earlier and continued to be hotly debated. Indeed, Planck himself originally promoted the wave nature of light and only came to his energy quantum description, as he states, out of "an act of despair … I was ready to sacrifice any of my previous convictions about physics" (Kragh 2000). His willingness to relinquish, however painfully, a preconceived notion of nature in favor of what his research was telling him is the hallmark of a great scientist. His surrender to the truth about the nature of light soon led to an entirely new branch of physics, quantum mechanics, and new insights into the inner workings of nature. Planck applied his quantum description of light to understanding the nature of blackbody radiation.

4.4 BLACKBODY RADIATION

All objects (including us!) have a temperature and emit some amount of electromagnetic radiation at all frequencies. Since we have evolved on a planet orbiting a yellow sun, our eyes are tuned to a relatively narrow band of frequencies which we parochially refer to as the "visible" spectrum. The visible spectrum includes what we perceive as color. Everyday experience tells us that an object can emit energy without us actually seeing it, e.g., the infrared heat we radiate. In the 1660s, Isaac Newton used a prism to disperse the light from the Sun into multiple colors, just as raindrops due to make a rainbow. In 1800, perhaps by making a connection between the color red in a rainbow and the red embers in a fire, the astronomer William Herschel performed an experiment where he placed a thermometer at the location in a projected spectrum of the Sun where the red light became too dim to be seen with the eye. At that location, he noted a rise in temperature. Herschel called this unseen light from the Sun "calorific rays", which we now refer to as infrared radiation (Herschel 1800). Unseen radiation from the Sun extends down the electromagnetic spectrum, from infrared to microwave, all the way to radio frequencies. Similarly, unseen radiation extends off the blue end of the solar spectrum, from ultraviolet to X-rays to gamma rays. Collectively, this broadband radiation from an object came to be referred to as blackbody radiation, a term which originates from the observation that truly black bodies absorb and emit all wavelengths of light. The term is something of a misnomer, since objects can have broadband spectral emission without necessarily being black. As discussed earlier, an object's perceived color is dependent on its temperature.

A mathematical description for the origin of blackbody radiation had been attempted by Balfour Stewart, Gustav Kirchoff, and Wilhelm Wien, all of which failed to fully match experimental data (Stewart 1858). Indeed, even Planck's first published attempt at deriving a blackbody radiation law was shown to be incorrect in the light of experimental data (pun intended). It was this failure that drove him to a quantum description of light (Eq. 4.1). Armed with this new insight, Planck was able to build on the work of Kirchoff, Wein, and others to develop Planck's law of blackbody radiation, Eq. (4.2).

$$B(\nu,T)=\frac{2h\nu^3}{c^2}\frac{1}{e^{h\nu/kT}-1} \tag{4.2}$$

where

$B(\nu,T)$ = intensity of a blackbody as a function of temperature and frequency (watts/(rad^2 m^2 Hz))

$\nu=\frac{c}{\lambda}$ = frequency (Hz)

h = Planck's constant (6.626×10^{-34} joule-sec)

k =Boltzmann's constant (1.38×10^{-23}joule/K)

c = speed of light (3×10^{8} m/sec)

T = temperature (K)

Planck's law (also referred to as the Planck function) allows one to calculate the specific intensity, I_ν, of a black body of temperature T at any frequency. A plot of the Planck function for surface temperatures characteristic of O, B, A, F, K, G, and M stars is shown in Figure 4.4.

The wavelength (or frequency) at which the Planck function reaches its peak is set only by the object's temperature and can be determined using Wein's Displacement Law,

$$\lambda_{peak} = \frac{2900}{T} \tag{4.3}$$

where

λ_{peak} = the wavelength at which the Planck function peaks (μm, with 1 μm = 10^{-6} m)

T = the temperature of the black body (K)

The total radiated power, or luminosity, of a blackbody at a given temperature is proportional to the area under the corresponding curve and can be found analytically by integrating the Planck function over frequency and multiplying by the object's surface area. For the case of spherical black bodies, *such as stars*, the resulting expression is,

$$L = 4\pi R^2 \sigma T^4 \tag{4.4}$$

where

L = radiated power of black body (watts)

R = radius of black body (m)

σ = Stefan-Boltzmann constant (5.67×10^{-8} watts/(m^2 K^4))

T= Temperature (K)

The above expression is called Stefan-Boltzmann's Law. It was derived by Josef Stefan in 1879 from experimental data published by John Tyndall in 1864 (Tyndall 1864) and from theoretical considerations of thermodynamics and Maxwell's equations of electromagnetism (Crepeau 2009). The expression is fully consistent with the later work of Planck (Stefan 1879; Boltzmann 1884). It relates three fundamental properties of black bodies: luminosity, temperature, and size. In the case of stars, the temperature can be inferred from their spectral class and the luminosity determined from knowledge of their apparent brightness and distance.

4.5 THE HERTZSPRUNG–RUSSELL (H–R) DIAGRAM

The next step in understanding the true nature of stars began by comparing the spectral classification and brightness of stars located at the same distance from Earth. This was first done in a paper by Hans Rosenberg in 1910, where he made an x–y plot of the apparent brightness of members of the Pleiades star cluster as a function of spectral type as determined from the strength of their hydrogen and calcium absorption lines (see Figure 4.5). Looking at the plot, the first thing he noticed was that the brightness of the stars was "a pure function" of their spectral type. Since members of the cluster are effectively at the same distance from the Earth, this meant that the power output, i.e., luminosity, of each

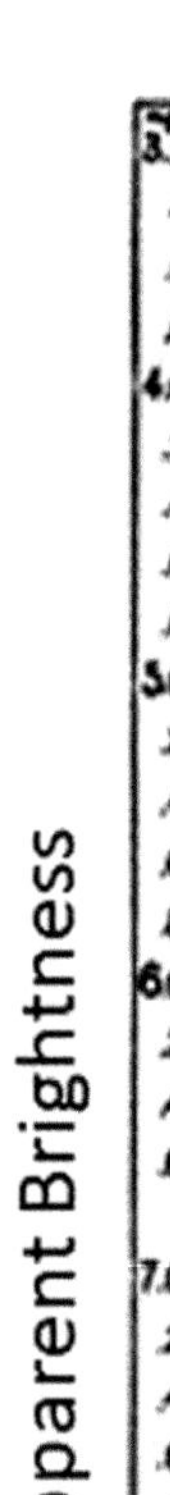

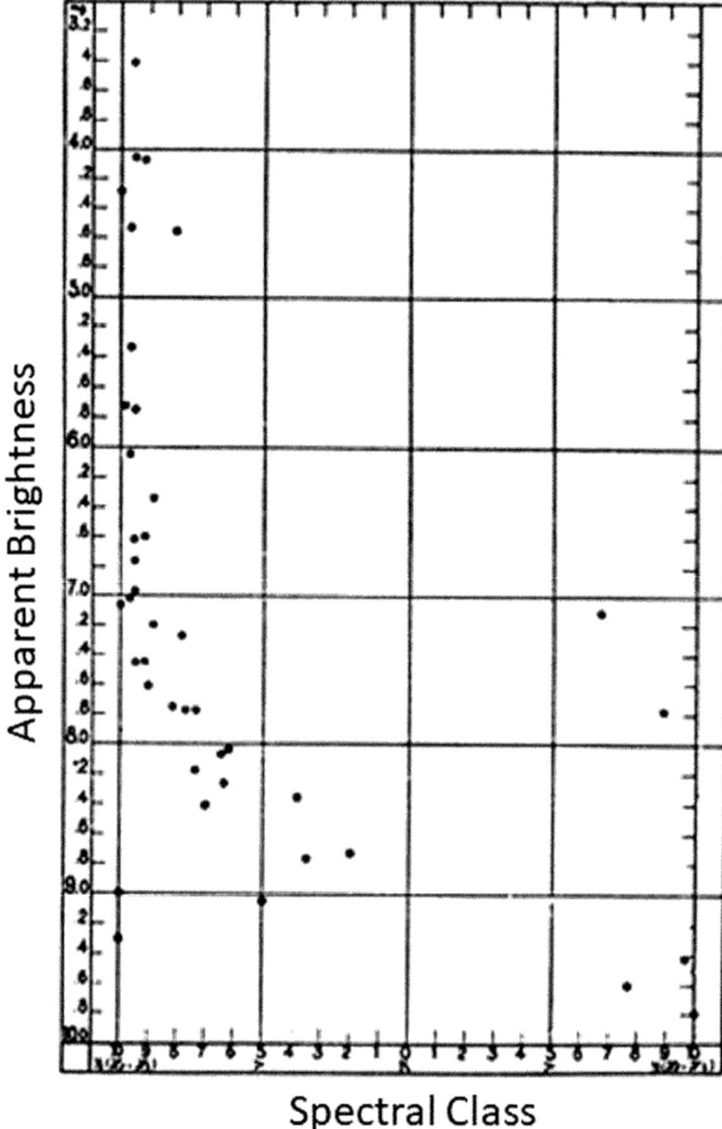

FIGURE 4.5 First plot of Stellar Brightness vs. Spectral Class. The dots represent members of the Pleiades star cluster. Since they are at approximately the same distance, their brightness can be compared directly. A clear correlation between brightness and spectral class can be seen. Image: adapted from Rosenberg 1910.

star was also a function of spectral type. Shortly after, a more well-known paper showing similar results for both the Pleiades and Hyades star clusters was published by Ejnar Hertzsprung (Hertzsprung 1911).

While observing stars in clusters provides a means of assessing the relative properties of stars within a cluster without knowledge of the actual distance to the cluster, the true absolute magnitude and therefore the luminosity of stars remains unknown. However, by the spring of 1913, the distances to several hundred stars had been determined using the

FIGURE 4.6 Henry Norris Russell (1877–1957) and family. He is credited for originating the mnemonic for spectral classification: "Oh Be A Fine Girl Kiss Me Now". His inspiration is clear to see. Image: ***ca.*** 1915.

parallax technique described in Figure 3.5. These distance determinations together with the associated stellar classifications allowed Henry Russell to make the first "modern" diagram of stellar absolute magnitude versus spectral type (Russell 1913; see Figure 4.6). This diagram, now referred to as the Hertzsprung–Russell or H–R diagram, will turn out to be as important for understanding stellar evolution as Hubble's diagram (Figure 3.14) was for understanding the evolution of the universe.

From the plot Russell reached several conclusions.

1) Stars of a given spectral class have, with few exceptions, a characteristic luminosity that varies by more than two orders of magnitude between classes; e.g., hot blue B stars are on average a 100× brighter than cooler A stars.

2) On the other hand, there are many red stars, such as Aldebaran, Arcturus, and Antares, that are as luminous as class A and/or B.

3) Stars appearing on the diagram as class K5 and M are either very bright or very faint; none are comparable to the Sun in brightness.

Russell also noted that the stars on the diagram "clustered principally close along two lines, one descending sharply along the diagonal, from B to M, the other starting also at B, but running almost horizontally". In his paper Russell noted that Hertzsprung had made a similar observation based on his study of star clusters and had designated stars that have, on average, a brightness ~100× that of the Sun as *giant* stars and all others as *dwarf* stars. An even more fundamental discovery was made by Russell concerning the mass of stars. Some of the stars in his sample were spectroscopic binaries. When such objects are observed over time a shift in the spectrum between the two-member stars can be observed. From this observed shift, it is possible to determine the orbital period and velocity of the two stars relative to one another. When combined with Kepler's Law of planetary motion, this information can be used to derive the mass of the stars. By comparing these mass estimates with luminosity, Russell correctly concluded that there is a correlation between the mass and brightness of a star, with the luminosity of a star increasing exponentially with mass.

Russell presented his results at a meeting of the Royal Astronomical Society in 1912 and later at a meeting of the American Association for the Advancement of Science in 1913. In attendance at the Royal Astronomical meeting was British physicist Arthur Stanley Eddington (see Figure 4.7). Russell's results motivated Eddington to apply his knowledge of thermodynamics, radiative transport, and fluid mechanics to develop a theory of stellar evolution that connected the dots in Russell's diagram. Eddington lays out his theory of stellar evolution in his classic 1926 paper "The Internal Constitution of the Stars" (Eddington 1926; Figure 4.8).

The greatest mystery of stellar evolution was the nature of the energy source behind a star's tremendous luminosity. Eddington concluded that chemical and gravitational energy were not sufficient to drive a star over its lifetime and speculated that the source of a star's power must be from energy released through fusing hydrogen nuclei (i.e., protons) into helium nuclei in a star's core. At normal temperatures and pressures protons repel one another due to their like positive charge. However, gravitational contraction within a stellar core drives temperatures to values in excess of several million degrees. The higher the temperature within a volume, the greater the kinetic energy of the particles contained in it. At such high temperatures the protons within a stellar core have sufficient kinetic energy to overcome the electrostatic repulsion force they feel toward one another and fuse them into helium nuclei. During the reaction, a small amount of proton mass is converted to energy through Einstein's iconic relation,

$$\Delta E = \Delta mc^2 \tag{4.5}$$

where

ΔE = energy released from nuclear reaction
Δm = amount of mass converted to energy
c = the speed of light

This was a remarkable intuitive leap on the part of Eddington, since the physics behind the transmutation of elements through thermonuclear fusion was yet to be developed. Indeed, in the paper he states, "But is it possible to admit that such a transmutation is occurring? It is difficult to assert, but perhaps more difficult to deny, that this is going on."

Eddington predicted that dwarf stars would have sufficient hydrogen in their cores to maintain a constant energy output over much of their lifetimes. This explained the large

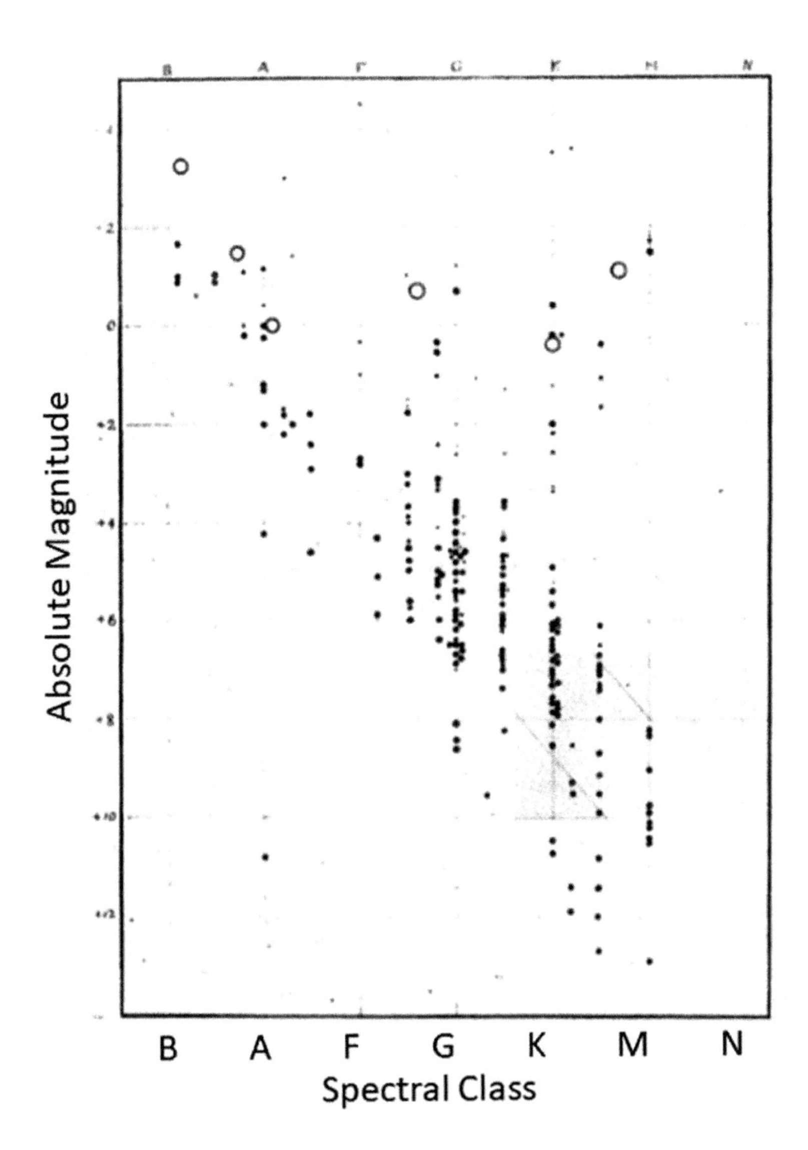

FIGURE 4.7 First Hertzsprung–Russell diagram. In the spring of 1913 Russell used all available parallactic distance measurements to compute the absolute magnitudes (i.e., luminosity) of ~300 stars. He then made this plot comparing stellar luminosities versus spectral classification for each star in his sample. This plot showed that the relationship between a star's luminosity and spectral classification was not random and motivated the pursuit of the underlying physics necessary to explain stellar evolution. Image: adapted from Russell 1913.

numbers of stars on the diagonal line on Russell's diagram, now referred to as the "main sequence". If a star is found to be on the main sequence, it can be inferred that it is burning hydrogen into helium within its core. During the course of its lifetime a star moves onto and off of the main sequence. Once Eddington asserted that nuclear fusion was the source of a star's energy, more detailed models of stellar evolution could be developed that traced the roller coaster ride through the H–R diagram a star follows over the course of its life. Eddington conveyed his bold ideas at some peril to his professional reputation. At the end

FIGURE 4.8 Arthur Stanley Eddington (1882–1944). Among Eddington's many contributions to astrophysics, perhaps the greatest is reaching the conclusion that stars derive their great power from nuclear energy. Image: ***ca.*** 1912.

of his paper, he compared himself to Icarus: "But if he is not yet destined to reach the Sun and solve for all time the riddle of its constitution, yet he may hope to learn from his journey some hints to build a better machine".

Even though they do not appear black to us, stars are able to absorb and emit radiation over a wide range of frequencies, which allows them to be classified as blackbody radiators. Since they are blackbodies, the relationship between their luminosity and temperature is subject to the Stefan-Boltzmann Law described by Eq. (4.4), which states a star's luminosity increases as the fourth power of its temperature and by the square of its radius. The H–R diagram plots the luminosity of an ensemble of stars as a function of their effective surface temperature. Using this information, together with the Stefan-Boltzmann Law, we can calculate the diameters of stars on an H–R diagram. Figure 4.9 is a modern version of Russell's H–R diagram and illustrates the wide variation in stellar size, luminosity, and temperature that exists in nature, all of which are governed by well-known physical laws. The fact that we can calculate the diameter of stars hundreds, thousands, or millions of light years away and understand the nature of their inner source of power is one of the many triumphs of the nineteenth- and twentieth-century astrophysics.

From examining Figures 4.7 and 4.9, it can be seen that almost all stars fall into five main categories as specified by their location: main sequence, giants, supergiants, white dwarfs, and red dwarfs. As deduced by Eddington, stars spend most of their lifetimes in the main sequence of stellar evolution, where they are burning hydrogen into helium within their cores. This is why most stars are found there. It is during their time on the main sequence that the energy output of a star is the most stable and likely to provide suitable conditions for the evolution of life within a stellar system.

As observed by Russell, the greater the mass, M, of a star, the greater is its luminosity, L. For main sequence stars the relationship between M and L is exponential (Salaris et. al. 2005).

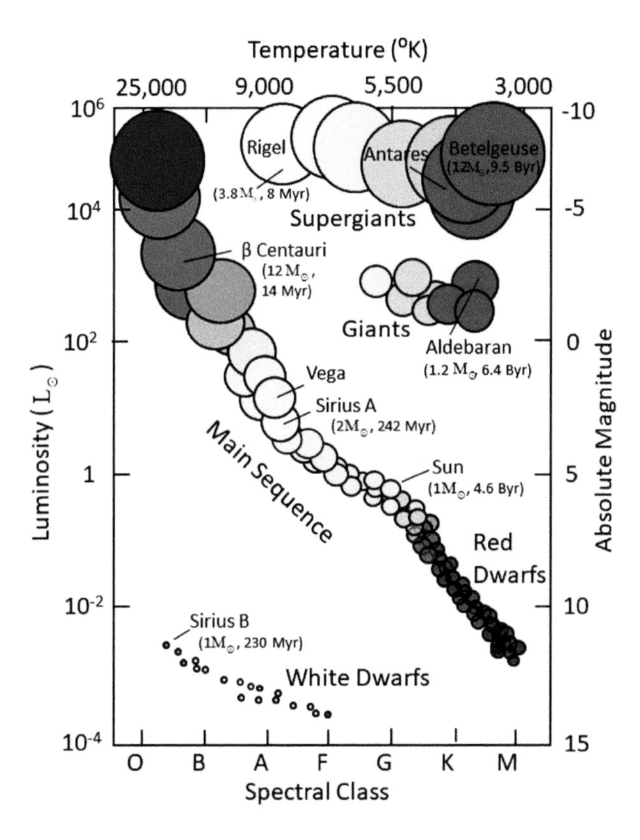

FIGURE 4.9 Modern H–R diagram. The H–R diagram is a scattergram of stars plotted as a function of their intrinsic luminosity versus their effective surface temperature. The stellar luminosity (y-axis) is typically expressed in terms of either solar luminosities (with $1M_{\odot}$being one solar luminosity) or absolute magnitude. The x-axis is expressed in terms of effective surface temperature or spectral class. About 90% of all stars can be found on the Main Sequence, where they are burning hydrogen into helium in their cores. All stars spend the majority of their lifetimes on the main sequence. Once stars like our Sun burn through the hydrogen in their cores, they quickly evolve off the main sequence; first becoming Giants, then Supergiants, and, ultimately, White Dwarfs, which account for 9% of the stellar population. The White Dwarf region of the H–R diagram is often referred to as the "stellar graveyard". Stars more than ~10× the mass of the Sun never make it there, instead ending their existence in a spectacular supernova.

$$
\begin{aligned}
L &\approx 0.23\left(\frac{M}{M_{\odot}}\right)^{2.3} \left(M < 0.43M_{\odot}\right) \\
L &= \left(\frac{M}{M_{\odot}}\right)^{4}, \left(0.43M_{\odot} < M < 2M_{\odot}\right) \\
L &\sim 1.5\left(\frac{M}{M_{\odot}}\right)^{3.5}, \left(2M_{\odot} < M < 20M_{\odot}\right)
\end{aligned}
\tag{4.6}
$$

where
L = stellar luminosity in solar luminosities ($L_\odot$)
M = stellar mass in solar masses ($M_\odot$)

The nonlinear relationship between the luminosity and mass of the star is due to the temperature sensitivity of nuclear reactions in its core. The greater the mass of the star, the greater will be the gravitationally induced pressure and subsequent temperature in its core. The hotter the core, the more often two protons will strike each other with sufficient momentum for them to fuse together and convert a small fraction of their mass into energy via Eq. (4.5). The more energy that is released the more luminous the star will appear. The more luminous a star is found to be, the faster it burns through its store of hydrogen in its core and the shorter will be its lifetime on the main sequence, τ_{MS},

$$\tau_{MS} \approx 10^{10}\left(\frac{M}{M_\odot}\right)^{-2.5}. \quad (4.7)$$

where
τ_{MS} = stellar lifetime on the main sequence (yrs).

Massive stars have the greatest luminosities and therefore have the shortest lifetimes, ~10 million years. The masses and lifetimes of several stars visible with the naked eye are shown in Figure 4.9.

4.6 THE LIVES OF STARS

Stars are in some ways like people; they are born, live their lives, and, when their bodies are exhausted, die. The destiny of a human is driven by a combination of genetics, personal choices, and environmental factors. The destiny of a star is driven by its mass.

4.6.1 The Birth of Stars

As described by Einstein's General Theory of Relativity, an object with mass will distort the space around it in such a way as to instill an attraction between it and other objects. The greater the mass, the greater the spatial distortion and the greater the attractive force. The strength of the attractive force decreases exponentially with distance between objects. For many applications this behavior can be adequately described by Isaac Newton's Law of Universal Gravitation.

$$F_{grav} = G\frac{m_1 m_2}{d^2} \quad (4.8)$$

where
F_{grav} = attractive gravitational force (N)
m_1 = mass of one object (kg)
m_2 = mass of another object (kg)

d = distance between the objects (m)
G = gravitational constant = 6.673 x 10^{-11} N m^2/kg^2

Eq. (4.8) holds whether the objects being considered are two protons, humans, planets, stars, or galaxies.

All stars form from the collapse of interstellar clouds of molecular hydrogen (H_2) that permeate much of the Milky Way (see Figure 1.1). These clouds are a mixture of diffuse and clumpy gas with a sprinkling of dust. They range in size and mass from ~1 ly and ~1$M_\odot$ to several 100 light years and ~$10^6 M_\odot$. The clouds are heated to ~10K by the light from stars and cosmic rays (i.e., high energy particles) from supernovae explosions. Within these gas clouds, there is a constant tug of war between thermal energy, which tries to disperse the cloud, and gravity, which tries to collapse it. The formation of a star (or stars) is triggered within a cloud clump when the mass density, ρ, of the clump is sufficiently high and the temperature, T, sufficiently low that the gravitational binding energy within the clump, U, is more than twice as much as the cloud's thermal kinetic energy, K. Under these conditions (known as the Virial Theorem), the clump will begin to collapse from the inside out (Shu 1977). If we assume the clump is spherical and has a uniform density, the mass of a clump under such conditions is referred to as the Jean's mass, M_J, (after the nineteenth-/early twentieth-century British astronomer James Jeans) and can be computed from the expression,

$$M_J = \left(\frac{5kT}{Gm}\right)^{\frac{3}{2}} \left(\frac{3}{4\pi\rho}\right)^{\frac{1}{2}} \tag{4.9}$$

where

M_J Jeans mass (g)
T= gas temperature (K)
ρ = gas mass density (g cm^{-3})
M = mass of H_2 = 3.3×10^{-24}gm
G = Gravitational constant = $6.674 \times 10^{-8} g^{-1}\ cm^3\ s^{-2}$
K = Boltzmann's constant = 1.38×10^{-16}erg K^{-1}

Examination of Eq. (4.9) indicates that for a given mass density, ρ, higher temperature clouds (e.g., the Orion molecular cloud) will preferably form higher-mass stars, while lower temperature clouds (e.g., Rho Ophiuchi) will preferably form low-mass stars. Here is why. As with any volume of gas, as a molecular cloud clump compresses, it will tend to heat up due to the higher frequency of collisions between its constituent molecules. Every collision will result in the transformation of some of the participating molecules' kinetic energy into heat, i.e., thermal energy. This added thermal energy will tend to drive up the value of M_J, because more gravitational energy is required to hold the clump together. However, if a collapsing clump is able to radiate away its heat, then the cloud will remain isothermal while its mass density, ρ, increases due to the infall of gas toward it center. This will have the effect of driving the value of M_J down, leading to the fragmentation of the clump into an ensemble of lower-mass clumps. The fragmentation will continue until the clumps reach a high enough density (~10^{10} H_2 molecules per

cm^3) that they trap their own heat (i.e., become optically thick to their own radiation). At which point the thermal energy within a clump, now referred to as a core, achieves equilibrium with its own gravitational energy and the collapse is halted. Such cores are classified as protostars and can be identified by their intense thermal radiation. A protostar is a young star that is yet to accrete the bulk of the mass it will have when it arrives on the main sequence.

4.6.2 Evolution of a One Solar Mass Star

As was briefly touched on in Chapter 1 and will be discussed further in later chapters, it took a long time for intelligent life to evolve on Earth. In fact, it took since the formation of the solar system from a collapsing molecular cloud clump about 4.6 billion years ago until now. If the evolution of life on Earth is typical of elsewhere, then Eq. (4.7) tells us that such life will most likely be found to be on (or to have originated from) planets orbiting stars with a mass similar to that of the Sun. Stars much more massive than the Sun ($1\,M_\odot = 2\times10^{33}\,gm$) will simply not be on the main sequence long enough for intelligent life to evolve. Stars significantly less massive than the Sun will be on the main sequence longer, but will, according to Eq. (4.6), radiate far less power (i.e., have too low a luminosity) to produce habitable conditions on the surfaces of their planetary offspring, except on those they hold very close. In terms of providing a nurturing environment for the evolution of communicable lifeforms, a relatively common G2 dwarf star like our own Sun is categorized as a being a "good" star. Due to their importance, let us look a bit more into their evolution.

4.6.2.1 Protostellar Evolution

Due to the shear forces present in molecular clouds resulting from galactic rotation, many molecular clumps are slowly rotating (~0.1 km/s). The outward centrifugal force produced by the rotation, together with thermal pressure, may just balance the clumps gravitational force resulting in a stalemate. In such a situation the clump is said to be in hydrostatic equilibrium. The fight against the inward pull of gravity could also be aided by turbulent energy within the clump and/or magnetic fields that are effectively "frozen" into a clump through interactions with ambient charged particles (i.e., ions). To ions the magnetic fields act like "rubber bands" that resist their passing. However, an overpressure in the surrounding molecular cloud, perhaps brought on by a passing supernova shockwave, may provide the extra push gravity needs to win the tug of war against outward pressure, such that the clump commences to collapse. In a typical molecular cloud, a clump may wait one to ten million years before conditions are right for collapse. In some instances, such conditions may never occur and the clump dissipates with the parent cloud before collapse occurs. The path of a collapsing solar mass molecular cloud core as it would appear on an H–R diagram is shown in Figure 4.10.

During the collapse process angular momentum is conserved, and the protostellar core spins up as it contracts. One important consequence of the rapid rotation is the formation of a disk (much like how pizza crust is made by spinning a ball of dough). Once formed, the young protostar continues to accrete material through its disk. The resulting shock at the boundary between the disk and infalling matter will produce copious amounts of

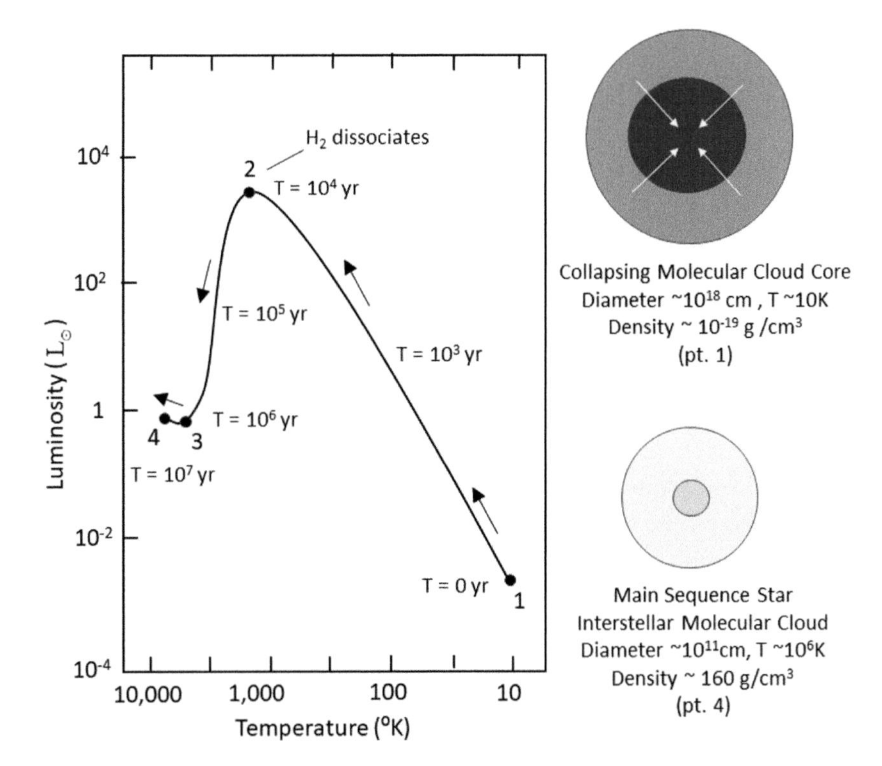

FIGURE 4.10 Protostellar evolution on the H–R diagram. A one solar mass protostar begins its life as an extended, ~10^{18} cm diameter, cold molecular cloud core on the bottom right of the extended H–R diagram (pt. 1). The core becomes gravitationally unstable and begins to collapse in on itself, causing the core to heat up and begin travelling up and to the left on the diagram. A young protostar forms at the center of the core. As material from the parent cloud continues to rain down on the protostar, an accretion disk forms around it, driving its luminosity up to ~1000x that of the Sun. When the protostar reaches ~2000 K the H_2 molecules begin to dissociate into protons (pt. 2). This is a strongly endothermic reaction, which serves as a thermal sink, allowing the protostar to further contract without heating up. This acts to drop the star's luminosity, sending it down the "Hayashi track" of the H–R diagram (between pts. 2 and 3). During this time, the interior of the star is fully convective, accretion slows, and planet formation commences (next chapter). Between pts. 3 and 4 the protostar is on the "Henyey track", where its interior further contracts, becomes partly radiative, and heats up to the point where its central core temperature reaches ~3 million °K, sufficient for the onset of nuclear fusion. The protostar is then classified as a main sequence star. Representative objects at each evolutionary phase.

ultraviolent (UV) photons that heats and ionizes the surrounding cloud material. A star's luminosity at this stage of evolution is 1000's of times brighter than when it is on the main sequence (point 2 in Figure 4.10; Zeilik 1979) and is due solely to the release of gravitational energy. As the disk accretes and contracts, it will continue to spin up. If the disk spins up too much, outward centrifugal force within the disk will prevent the inward flow of material onto the protostar itself. In an effort to dump rotational angular momentum from the disk back to the parent cloud, the protostar forms jets of gas that shoot out from its north and south poles at ~100 km/s. These bipolar jets deposit ~10% of the inward gas mass, and its associated angular momentum, back into the parent cloud. The jets can extend into space for several light years (Snell et al. 1980; Shu et al. 2000). With the angular momentum blockade gone, the protostar can continue to accrete material from the parent cloud through the disk until the supply of surrounding gas is exhausted or other processes within the accretion disk prevent its transfer. As accretion subsides, so will the bipolar outflow.

What was once a hot (~1,000 K) accretion disk will cool and become a protoplanetary disk (or proplyd, for short). After the cessation of accretion, the protostar will continue to collapse internally under its own weight until its central temperature reaches ~3,000,000 °K, at which point nuclear fusion is triggered in its core and the protostar officially arrives on the main sequence, becoming a zero age, main sequence (ZAMS) star. Even before arriving on the main sequence, dynamical process within the protoplanetary disk can lead to the formation of planets, moons, asteroids, and/or comets (see Chapter 5). The entire star formation process, from being an interstellar molecular cloud clump with a gas density far less than that of the air you breathe, to being a solar mass star with a young planetary system, takes only ~10 million years; about 0.1% of the star's main sequence lifetime (see Figure 4.11).

Where exactly a star will end up on the main sequence depends primarily on its mass, with higher mass stars on the upper left and lower mass stars on the lower right (see Figure 4.9). While on the main sequence the star is fusing hydrogen into helium in its core. Hydrogen is the fuel the star burns and helium is the ash left behind. In the case of a nuclear bomb the initiation of a fusion reaction leads to an outward explosion. In the case of a main sequence star, the outward pressure generated from the heat of the nuclear reaction in its core is held in check by the inward pressure exerted by gravity trying to pull everything toward the center. When the tug of war between a star's gravitational energy and thermal energy is a stalemate, the star is said to be in hydrostatic equilibrium. A star will remain on the main sequence until a significant fraction of the hydrogen in its core is consumed. For a solar mass star, this will not happen for ~10 billion years (Eq. 4.7). During this time the size and energy output of the star will remain remarkably constant. It will patiently provide a nurturing environment over the eons of time required for the possible evolution of intelligent life on its planetary offspring.

4.6.2.2 Evolution off the Main Sequence

Our Sun is 4.6 billion years old, which means it has about 5 billion years to go before it runs out of hydrogen to burn in its core. The universe is 13.77 billion years old, which means there already have been solar mass stars that have passed the 10 billion year evolutionary milestone.

What happened to them, not to mention the planets that orbit them? *Alien civilizations could have existed on these worlds millions, if not billions, of years before our Sun was even born.*

The post main sequence evolution of a solar mass star can be plotted on an H–R diagram (see Figure 4.12). Once a star burns through the hydrogen in its core, nuclear reactions will cease and it will begin to cool (pt. 5 on Figure 4.12). Gravity gains the upper hand in its tug of war with thermal energy and the core begins to contract. As the core contracts

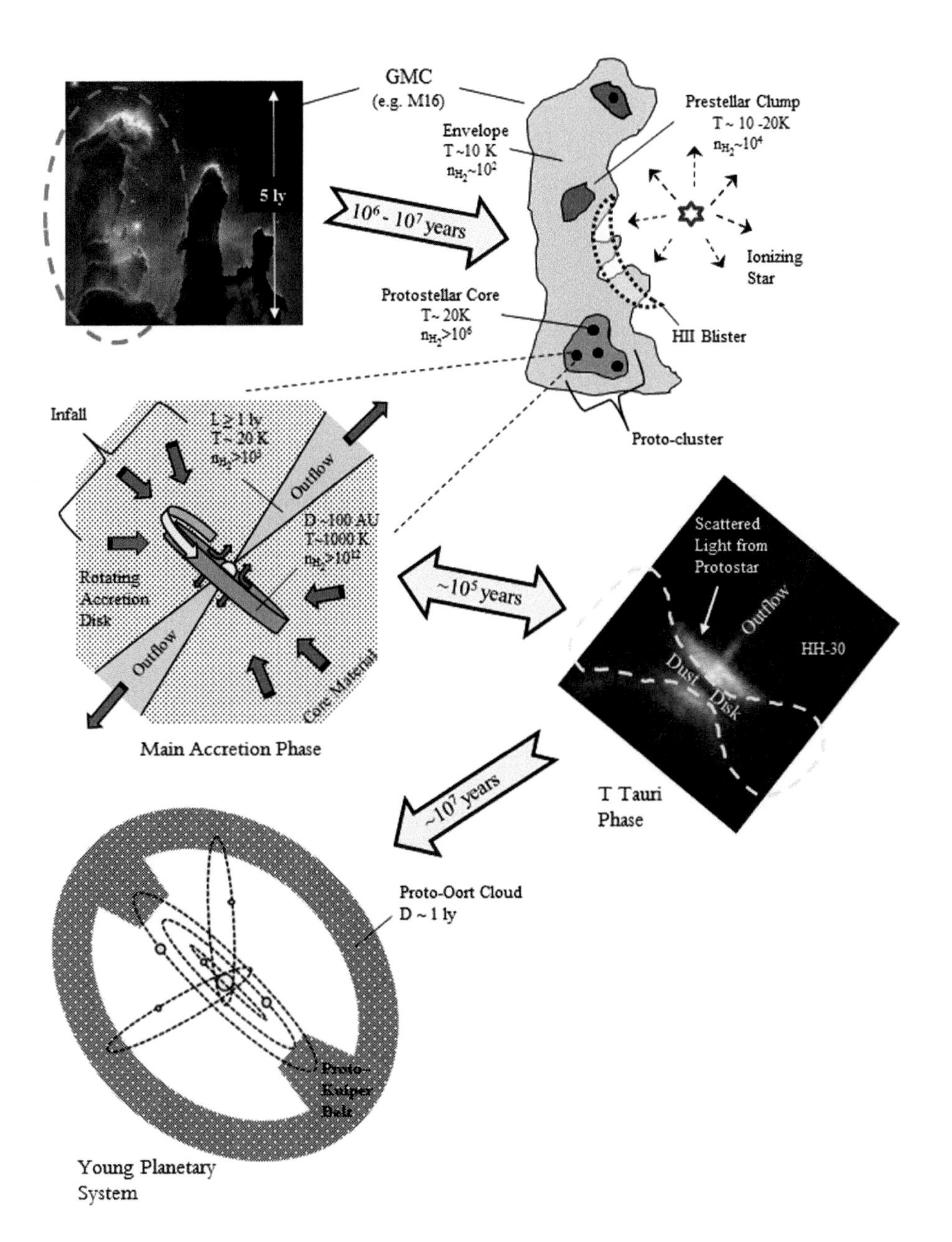

FIGURE 4.11 Stages of protostellar evolution. In just ~10^7 years a young planetary system can form from the collapse of a clump within a Giant Molecular Cloud (GMC), providing an oasis for life in a vast cosmic desert. Surrounding the system at a distance of ~1 ly is a cloud (i.e., the Oort Cloud) of cometary debris left over from the planet formation process. Occasionally gravitational perturbations will send one of these comets into the inner solar system, providing a fossil record of the primordial solar system for Earth-bound observers. Image credits: M16: NASA, Jeff Hester, and Paul Scowen; HH-30: NASA Hubble; Fomalhaut Protoplanetary system: David A. Hardy.

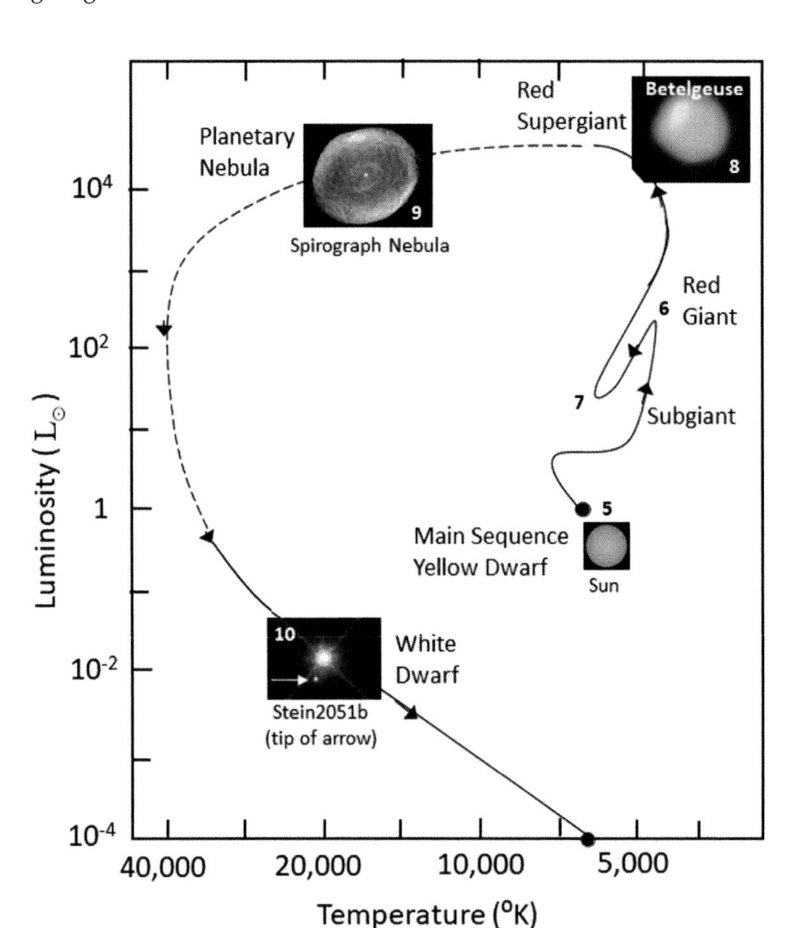

FIGURE 4.12 Post main sequence evolution of a solar mass star. Once a Sun-like star fuses all the hydrogen in its core to helium (pt. 5), the core contracts and heats up, igniting the shell of hydrogen that surrounds it. The heat from the shell causes the outer layers of the star to expand, thereby driving it up the H–R diagram, while the inner core continues to contract. Once the core reaches ~100 million °K, there is a "helium flash" (pt. 6) and helium begins to burn into carbon and oxygen in the core (pt. 7). Burning continues until all the helium in the core is burned, at which time the core once again contracts and ignites another round of shell burning. However, this time it is helium ash from the hydrogen-burning shell that is consumed. Now with two concentric burning shells, the star rapidly expands and continues its climb up the H–R diagram until it becomes a supergiant star (pt. 8). The temperature sensitivity of the helium-burning shell, together with the low surface gravity resulting from being so extended, make the outer layers of the star unstable. It begins to pulsate. Over time the pulsations become so violent that the outer layers of the star are ripped off, revealing the hot carbon-oxygen core. At this point, the star has become a planetary nebula, with the exposed core becoming the center star and the outer layers expanding out into the interstellar medium (pt. 9). The core continues to collapse until its constituent electrons have a velocity approaching that of light. The core is now classified as a white dwarf. Overtime it will cool from blue to red to black and continue to exist for up to 10^{200} years (essentially forever). The time between when the star leaves the main sequence until it becomes a white dwarf is ~1.5 billion years. Image credits: Sun (NASA); Betelgeuse (ALMA (ESO/NAOJ/NRAO)/E. O'GORMAN/P. KERVELLA); Spirograph Nebula (NASA, ESA, and the Hubble Heritage Team (STScI/AURA)); Stein2051b (NASA, ESA, and K. Sahu (STScI); Data from Shu 1982.

it heats up and ignites the shell of unburned hydrogen enveloping it. Due to the elevated temperatures within the contracting core, the shell burns even more fiercely than the core had previously. The shell burns outwards leaving a wake of helium ash behind it, in a manner similar to that of a fire raging through a forest. As before, the gravitational pressure from overlying gas layers keeps the nuclear fire from advancing too rapidly. However, the heat from the advancing burning shell causes the star's outer gas layers to expand and cool. As the layers expand, the diameter of the star increases and its surface temperature drops, even as its luminosity increases (see Eq. 4.4). These conditions cause the star to move off the main sequence and up and to the right on the H–R diagram (pts. 5–6). What was once a Sun-like star will in ~500 million years become a red giant star, with a surface temperature of 4,000 K and diameter 400× that of the Sun. Due to its immense size, its luminosity will be ~10,000× what the Sun's is now. When it reaches this stage the Sun will engulf Mercury, Venus, and the Earth. An example of one such star observable with the naked eye is Aldebaran (see Figure 4.9), the "eye of the bull" in the constellation of Taurus.

Since leaving the main sequence the core of the star continues to contract and heat up, the situation exacerbated by the mass of helium ash added to it by the hydrogen-burning shell above it. This continues until the core reaches a sufficiently high temperature (~100 million °K) to burn its helium ash into carbon and oxygen via the triple-alpha process. This occurs ~1.2 billion years after leaving the main sequence. The rush of energy causes the helium core to expand and cool, reducing the temperature at the base of the hydrogen-burning shell, thereby reducing the rate of nuclear reaction and dropping the star's luminosity and size. This causes the star to dip down to the right on the H–R diagram, at which point it is classified as a horizontal branch star (pt. 7). The star will remain here, burning helium into carbon and oxygen in its core and hydrogen into helium in a shell for ~50 million years, at which time the helium in the core is exhausted. Once the nuclear fire in the core goes out, it contracts. As before, the contraction will drive the temperatures just outside the core to a sufficient height to initiate burning in the surrounding layer of the ash. Unlike before, it is a layer of helium ash that is ignited. The star now contains two burning shells: an outer shell of burning hydrogen that leaves helium in its wake and an inner burning shell that consumes the helium ash left behind by the first (see Figure 4.13). The addition of a second burning shell drives another epoch of expansion and the star ascends the giant branch once again, now becoming a supergiant (pt. 8). It will remain a supergiant for ~500 million years. Betelgeuse in Orion and Antares in Scorpio are examples of such stars (see Figure 4.12).

The rate of the triple-alpha reaction in the helium-burning shell is extremely sensitive to temperature. This sensitivity, together with the star's low surface gravity due to it being so extended, makes it unstable. Small variations in temperature can cause the star to pulsate rapidly. Over time the pulsations grow larger. Once the pulsations become violent enough, the outer layers of the star will reach escape velocity and the star will shed its outer layers, exposing its carbon core. The expelled layers form a planetary nebula and the hot core becomes the central star (pt. 9, see Figure 4.13). The layers continue to expand. As they expand the layers thin-out until they fade into the background of the interstellar medium. Meanwhile, the leftover core contracts and heats up but never gets hot enough to burn

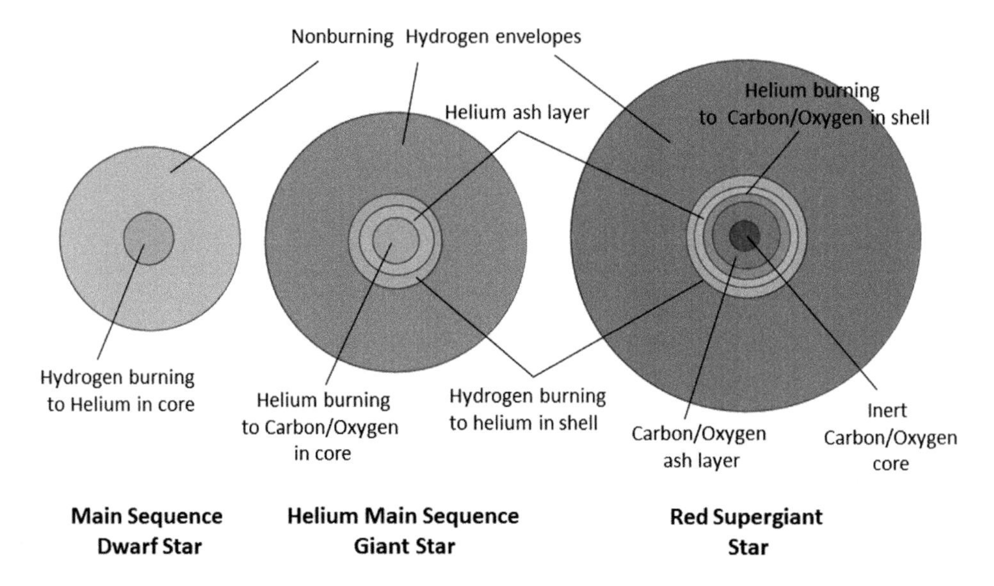

FIGURE 4.13 Cross sections of a solar mass star. As a solar mass star takes its roller coaster ride through the H–R diagram (Figures 4.10 and 4.12), its internal structure undergoes radical changes as it responds to the tug of war between gravity and thermal energy that dominates its existence.

carbon. The core contraction continues for ~75,000 years until the electrons within it attain velocities approaching the speed of light (the cosmic speed limit, thank you Einstein), at which time the contraction must stop. The carbon core is now classified as a white dwarf star (pt. 10). It will slowly cool and change colors from blue to red to black. If left undisturbed, it may remain a black dwarf star for as long as protons exits, which could be as long as 10^{200} years! This will be the fate of our Sun. The nearest known white dwarf is Sirius B, part of the Sirius binary star system, located 8.6 ly from Earth.

What is the fate of the planets orbiting a solar mass star? As mentioned earlier, the inner ones will be consumed by the star as it swells to become a red giant. If there are outer planets, they will likely survive and continue to orbit their parent through cosmic time. Several such enduring planetary systems have been found by space observatories. These include WD J0914+1914 (Gansicke et al. 2019) and WD 0806-991 (Luhman et al. 2011). These systems are located 2,038 and 63 ly from Earth, respectively.

4.6.3 Evolution of High-Mass Stars

If high-mass stars do not live long enough (i.e., billions of years) to provide a suitable environment for nurturing intelligent life on its planets, why should we concern ourselves with them? We are composed of oxygen, hydrogen, nitrogen, carbon, calcium, and phosphorus, with trace amounts of sulfur, potassium, sodium, chlorine, and magnesium. We also live on a planet made mostly of iron, oxygen, silicon, and magnesium. Just after the Big Bang the universe contained only the three lightest elements: hydrogen, helium, and lithium. So, where did all the heavier elements come from? *They were fused together in the centers of high-mass stars.* Toward the end of their days low--mass stars like our Sun will produce carbon and oxygen and become carbon–oxygen white dwarfs. Stars with masses significant less than that of our Sun will be unable to fuse helium and will become helium white dwarfs. Stars between 8 and 10.5$M_\odot$will achieve core temperatures hot enough to fuse carbon, but not much else and become oxygen-neon-magnesium white dwarfs. But except for relatively small amounts of gas released during the formation of planetary nebulae, stars < 10.5$M_\odot$will take these elements with them to their cosmic graves. So, it is somehow up to the high-mass stars to seed the universe with the heavy elements required to make planets and lifeforms. There are far fewer high-mass stars than low-mass stars, similar to the fact that there are far more people of modest means than there are rich people. Also, due to the mass–luminosity relation (Eq. 4.6), it is the massive stars that dominate our view of the night sky, just as rich people seem to dominate the evening news. However, the notoriety of high mass stars comes at a cost, the mass that makes them so luminous will also lead them to an early demise.

As in the case of low-mass stars, high-mass stars form from the collapse of a portion of a molecular cloud. However, the clump from which they originate will be both more massive and hotter than the nascent clump of their low-mass kin. The collapse of the clump is also more likely to have been triggered by the passing of a shockwave, rather than occurring spontaneously after a long prenatal period. Due to the greater pull of gravity, temperatures and density increase rapidly within the collapsing clump, to the point that nuclear fusion is initiated even as material is still raining in. The outward radiation pressure (in the form of light and heat) will work to slow or reverse the infall. One idea for how this turnaround can be averted is for accretion to occur through a disk as it does in low-mass stars. Further evidence for disk accretion is the observation of bipolar molecular outflows from high-mass stars similar to, but more powerful, than those seen emanating from the rotating disks of low-mass stars (MPIA 2021). Unlike low-mass stars, high-mass stars do not go through a higher-than-normal luminosity phase during their protostellar evolution, but evolve rapidly to the right on the H–R diagram, taking their place, however briefly, on the main sequence (see Figure 4.14).

How long a high-mass star stays on the main sequence is, as we know, a strong function of its mass. For example, a 5 $M_\odot$star will burn hydrogen into helium in its core for ~180 million years, while a 25$M_\odot$star will do so for only ~3 million years (per Eq. 4.7). Once the hydrogen is exhausted, the core will contract, heat up, and ignite the shell of fresh hydrogen surrounding it. Just as in the case of lower-mass stars, whenever a shell is ignited, the star will begin expanding; the outer layers cool off, moving the star to the right on the H–R diagram (see Figure 4.15). But unlike the case of low-mass stars, its luminosity does not increase significantly. Also, as before, as the shell is driving outward in the star, the core continues to collapse, which will lead to the ignition of the ash left behind by the burning shell. In the low-mass stars, there was first a hydrogen-burning shell and then a helium-burning shell. The temperatures within a low-mass star never reach the point where they are high enough to burn the carbon ash leftover from burning helium. As noted above, higher-mass star can burn heavier elements. In fact, a high-mass star, say 15 $M_\odot$, will go through four more cycles of core exhaustion, contraction, and subsequent shell ignition until there are six concentric shells burning simultaneously in the star, each one fusing together an element heavier than the last. To the outside observer, the star appears

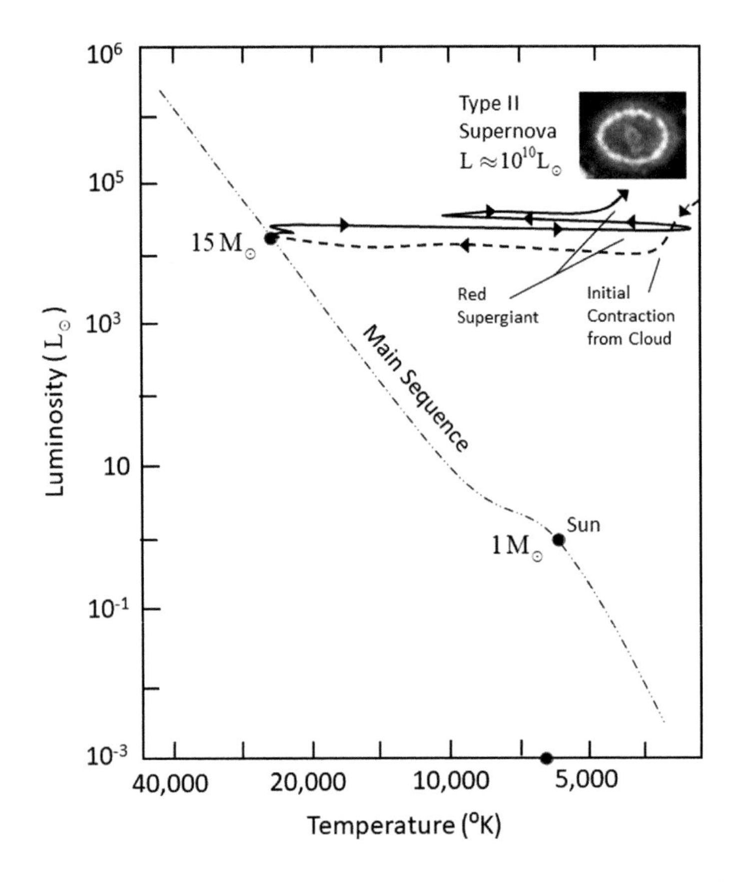

FIGURE 4.14 High-mass stellar evolution on the H–R diagram. A high mass, $15M_{\odot}$ star begins its life on the top-right of the H–R diagram, where it rapidly contracts from a high-mass molecular cloud core. The star rapidly moves cross the H–R diagram and takes its place on the main sequence in just ~60,000 years. After ~11 million years (see Eq. 4.7), it will burn all the hydrogen in its core into helium. As in the case of it low-mass kin, the core will begin to collapse and heat up. The temperatures will rise in the core until fusion is triggered in the as yet unburned shell of hydrogen that surrounds it. The star will then rapidly expand and move back to the righthand side of the diagram where temperatures at the core boundary are high enough to burn the helium ash dropped there by the expanding hydrogen shell. This contraction, shell ignition cycle will happen four more times until there are a total of six concentric, outward burning shells in the star, the ash of each one providing the fuel for the one below. At the center of the star is an inert iron-56 core, which will never get hot enough to burn. Once achieving this internal structure (see Figure 4.15), the star becomes unstable to gravitational collapse and will soon explode in a supernova Type II explosion. An example of a recent supernovae explosion is SN 1987A. Image credit: (NASA, ESA -J. Hester and A. Loll).

to zigzag back and forth across the top of the H–R diagram. The star does this in an unconscious, desperate attempt to maintain hydrostatic equilibrium. Each time it runs through the energy provided by fusing one element, it has to heat up enough to burn and release the nuclear energy contained in the next. This works fine until the star tries to fuse iron-56. Trying to add an additional proton or neutron to an iron-56 nucleus will actually reduce its binding energy, making it more likely to come apart than stay together. In other words,

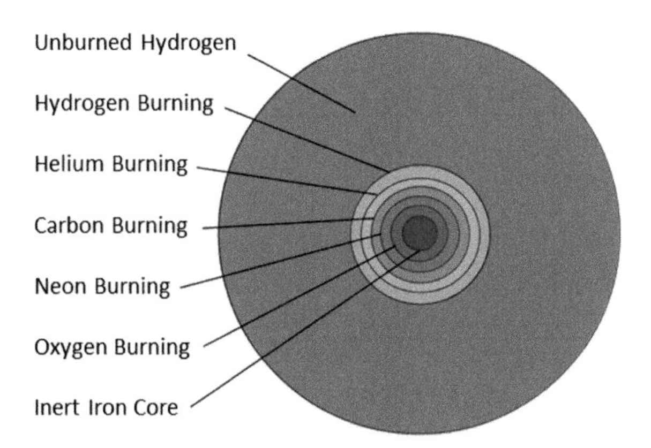

FIGURE 4.15 Internal structure of a presupernova high-mass star. In desperate attempts to maintain hydrostatic equilibrium, a high-mass star will burn through all its nuclear energy reserves, leaving it with six nested, burning shells. Each shell is fusing the material from the shell above it into a heavier element, that will then feed the shell below. The star is eating itself up from the inside out. At the center of it all is an inert iron-56 core, which can neither participate in nuclear fusion or fission. Having no central source of energy, the star will soon collapse in on itself, detonating a supernova explosion. Figure not to scale.

there is no release of energy from a fusion reaction with iron-56. It takes more energy to burn it than you can get out of it. It is analogous to trying to light a wet log with a match. In terms of nuclear reactions, iron is the pivotal element in the Periodic Table. Elements lighter than iron can release energy through nuclear fusion. Elements heavier than iron (e.g., uranium) can release energy through nuclear fusion. Iron is a nuclear no man's land. Once the high-mass star burns all the silicon in its core to iron, the nuclear fire goes out, and there is not sufficient thermal energy to balance gravity. The star's core collapses in on itself, creating temperatures up to 100 billion degrees. In stars less than about $20M_{\odot}$ the inner part of the core is compressed to neutrons. The neutron core is so dense that infalling material bounces off of it, creating an outward-bound shock wave that rips the star apart, creating a supernova explosion and leaving behind a core composed of neutrons. The isolated core is classified as a neutron star (Fryer 2006). When the progenitor star is greater than about $20M_{\odot}$ nothing can stop the collapse, not even the density of neutrons. In this case the inward rush of material drives densities up so high that the fabric of space and time can no longer support the core. The fabric rips and a black hole is formed (Fryer 1999). The radius, r_s, of a black hole is proportional to the mass of the star that created it, the expression for which was first derived in 1916 by Martin Schwarzschild in his solution to Einstein's General Relativity field equations (Schwarzschild 1916).

$$r_s = \frac{2GM}{c^2} \tag{4.10}$$

where

r_s = the Schwarzschild radius of a black hole
M = the mass of the progenitor object (e.g., star)

G = the gravitational constant
C = the speed of light

Black holes created by the collapse of a star are spherical and typically have a diameter of less than ~ 10 miles.

The temperatures and densities within a supernova explosion are sufficient to fuse together the remaining 92 naturally occurring elements in the Periodic Table (see Appendix 2). During its peak, a supernova explosion can outshine the whole galaxy in which it resides. A spherical shock wave emanates outward from the location of the supernova at velocities >1,000 km/s, seeding the galaxy with a treasure trove of heavy elements created by the star during its lifetime (a fine legacy). As the shock wave passes through the interstellar medium it sweeps up material in its path, creating large-scale supernova remnants (see Figure 4.16). In sweeping up material along its way, the shock wave loses momentum and slows down. After thousands of years its velocity will slow to the point where it becomes indistinguishable from the background turbulence of the interstellar medium. The passage of supernova shock waves through nearby molecular clouds can trigger the collapse and formation of a new generation of stars, completing the life cycle of stellar evolution.

Supernovae of the kind just described are classified as Type II supernovae. A Type I supernovae occurs in a binary system where one of the members is a white dwarf. For a supernova to occur, the white dwarf must first siphon off enough of its companion's mass to create the required pressure and temperature within its own core needed to burn carbon, thereby creating a runaway thermonuclear reaction that culminates in a supernova explosion. Type I supernovae occur in the Milky Way on average once every 50 years. Fourteen Type II supernovae have been observed in the Milky Way over the last two millennia (Figure 4.16).

SUMMARY

In this chapter, we have investigated how humankind used observations of the night sky, together with deductive reasoning and experiment, to understand the true nature of stars. Without stars, there would be no heavy elements for planets or a steady source of energy to nurture life on planetary surfaces over the billions of years required for the emergence of intelligent life. The understanding of stellar evolution is one of humanity's greatest achievements, and it did not come easy. It took the collective work of ~250 generations of observers, mathematicians, physicists, and technologists to gain the insights we have today. The great advances achieved in astrophysics over the last century came in the wake of technological leaps in telescopes and instrumentation, together with insights into quantum mechanics and the nature of space and time itself. We now know that it is in the vicinity of stars like our Sun that conditions will be the most conducive for harboring planets with advanced life. We also know that protoplanetary disks are a natural consequence of the star formation process. In the next chapter, we will investigate how planets form and which types are expected to be best for life.

FIGURE 4.16 Supernova Remnant. The Crab Nebula is the remnant of a supernovae explosion that took place in the constellation of Taurus in 1054 AD. Being "only" 6,500 ly away. It was sufficiently bright to be seen by Chinese, Japanese, and Native American astronomers, who noted its appearance. The explosion was so bright it could be seen during the day for several weeks. Even now, the nebula is being driven outward at a velocity of 1,500 km/s by the shock wave from the supernova. At the center of the nebula is a neutron star, the remnant of the progenitor star's stellar core. The neutron star is rotating on its axis 30 times a second. Magnetic fields trapped in the star during its collapse spin with it. Trapped electrons are whirling around the magnetic field lines at nearly the speed of light. As they do so, the acceleration of the electrons cause them to emit synchrotron radiation which is responsible for the blue glow. The magnetic fields are aligned with the rotation axis of the star and produce beams of radiation from the star's north and south poles, which are observed as electromagnetic pulses (from X-Ray to radio) each time the beams sweep by the Earth, like the rotating beams of a lighthouse. This type of object is called a pulsar and can be found throughout the Milky Way and beyond.

PROBLEMS

1) Compute the energy of photons at wavelengths of 0.1, 1, 10, 100, and 1,000 microns.

   Which wavelengths correspond to UV, optical, infrared, and terahertz photons?

2) At what wavelengths do F, G, K, and M stars radiate most of their photons?

3) What is the gravitational force between you and the Earth? Between the Earth and the Moon?

4) What is the required mass density for a 1 solar mass cloud to collapse if its average temperature is 10, 50, or 100 K? How do these densities compare to what is in your room?

5) When is a star like our Sun the brightest?

6) Compute the main sequence lifetime of a 0.2, 0.5, 1, 5, and 10 solar mass star.

7) What percent of a star's lifetime is spent in the protostellar phase?

8) What element cannot participate in a nuclear fusion reaction? Why?

9) Why do very high-mass stars go supernova?

10) About how many times have the atoms in your body participated in a supernova explosion?

11) If our Sun was to become a black hole (which it will not), what would be its diameter?

REFERENCES

Boltzmann, L. 1884. "Ableitung des Stefan'schen Gesetzes, Betreffend die Abhängigkeit der Wärmestrahlung von der Temperatur aus der Electromagnetischen Lichttheorie" ["Derivation of Stefan's Law, Concerning the Dependency of Heat Radiation on Temperature, from the Electromagnetic Theory of Light."] *Annalen Der Physik & Chemie (in German)* 258, no. 6: 291–4.

Crepeau, J. 2009. "A Brief History of the T⁴ Law." In *Proceedings HT 2009 ASME Summer Heat Transfer Conference*, San Francisco, July 19–23.

Eddington, A. 1926. *The Internal Constitution of Stars*. Cambridge University Press. ISBN 0-521-33708-9.

Fryer, C. L. 1999. "Mass Limits for Black Hole Formation." *Astrophysical Journal* 522, no. 1: 413–8.

Fryer, C. L., and K. C. B. New. 2006. "Gravitational Waves from Gravitational Collapse." *Max Planck Institute for Gravitational Physics*. Archived from the original on December 13, 2006. Accessed December 14, 2006.

Gänsicke, B. T., M. R. Schreiber, O. Toloza, F. Gentile, P. Nicola, D. Koester, and C. J. Manser. 2019. Accretion of a Giant Planet onto a White Dwarf. Parkinson's Disease Foundation. *ESO*. Archived (PDF) from the Original on December 4, 2019.

Kragh, H. 2000. "Max Planck: The Reluctant Revolutionary." *Physics World*. Archived November 5, 2018 at the Wayback Machine.

Herschel, W. 1800. "Experiments on the Refrangibility of the Invisible Rays of the Sun." *Philosophical Transactions of the Royal Society of London* 90: 284–29.

Hertzsprung, E. 1911. "On the Use of Photographic Effective Wavelengths for the Determination of Color Equivalents." *Publications of the Astrophysical Observatory in Potsdam* 1(63): 22.

Jones, A. R. 2021. "Hipparchus." *Encyclopedia Britannica*, February 10, 2021. Accessed September 22, 2021. https://www.britannica.com/biography/Hipparchus-Greek-astronomer.

Luhman, K. L., Adam J. Burgasser, and J. J. Bochanski. 2011. "Discovery of a Candidate for the Coolest Known Brown Dwarf." Astrophysical Journal Letters 730, no. 1: L9.

Mark, J. 2022. "Hipparchus of Nicea." March. https://www.worldhistory.org/Hipparchus_of_Nicea/.

Rosenberg, H. 1910. "Über Den Zusammenhang von Helligkeit und Spektraltypus in den Plejaden." *Astronomische Nachrichten* 186, no. 5: 71–8. https://www.leosondra.cz/en/first-hr-diagram/.

Russell, H. N. 1913. "Relations Between the Spectra and Other Characteristics of the Stars." *Popular Astronomy* 22: 275–29.
Schwarzschild, K. 1916. "Über das Gravitationsfeld eines Massenpunktes nach der Einsteinschen Theorie." *Sitzungsberichte der Deutschen Akademie der Wissenschaften zu Berlin, Klasse fur Mathematik, Physik, und Technik*: p. 189.
Shu, F. 1982. *The Physical Universe*. Mill Valley: University Science Books.
Shu, F., Najita, J., Shang, H., and Li, Z., 2000, "X-Winds Theory and Observations", in V. Mannings, A.P. Boss, S. S. Russell eds. *Protostars and Planets IV*. Tucson: University of Arizona Press, p. 789–814.
Snell, R., Loren, R., and Plambeck, R., 1980. "Observations of CO in L1551: evidence for stellar wind driven shocks", ApJ, 239, L125.
Stefan, J. 1879. "Über die Beziehung Zwischen der Wärmestrahlung und der Temperatur" ["On the Relation Between Heat Radiation and Temperature."] *Sitzungsberichte der Kaiserlichen Akademie der Wissenschaften: Mathematisch-Naturwissenschaftliche Classe (Proceedings of the Imperial Philosophical Academy [of Vienna]: Mathematical and Scientific Class)* (in German) 79: 391–428.
Stewart, B. 1858. "An Account of Some Experiments on Radiant Heat." Transactions of the Royal Society of Edinburgh 22: 1–20.
Toomer, G. J. 1988. "Hipparchus and Babylonian Astronomy". In *A Scientific Humanist: Studies in Memory of Abraham Sachs*, edited by Erle Leichty, Maria de J. Ellis, and Pamel Gerardi, vol. 9. Philadelphia: Occasional Publications of the Samuel Noah Kramer Fund.
Tyndall, J. 1864. "On Luminous [I.E., Visible] and Obscure [I.E., Infrared] Radiation." *Philosophical Magazine*. 4th Series 28: 329–341, see 333.
von Spaeth, O. 2000. "Dating the Oldest Egyptian Star Map." *Centaurus* 42, no. 3: 159–79.
Zeilik, M. 1979. *Astronomy: The Evolving Universe*. New York: Harper and Row.

CHAPTER 5

Planet Evolution

PROLOGUE

Time is woven into the fabric of the universe. The inexorable forward motion of time forces everything in the universe to evolve. The degree to which evolutionary changes are manifested within a particular object from moment to moment varies. The characteristics of the universe itself changed rapidly at the beginning and have since "settled down" into an everlasting epoch of expansion and cooling. Stars like our Sun also evolve, going through rapid changes during formation and then settling down to long midlife before entering a perpetual cooling phase. Planets too evolve. They form rapidly as a consequence of star formation and, if they achieve a stable orbit about a Sun-like star, will exist for billions of years. If conditions are suitable on (or in) a planet, life itself may form and evolve.

5.1 INTRODUCTION

The existence of extrasolar planets orbiting Sun-like stars has been theorized since the times of the ancient Greeks (see Section 1.2.3). Exactly how such planetary systems formed became the subject of debate among eighteenth-century philosophers and scientists. One of the first attempts to describe the process was by the Swedish philosopher Emanuel Swedenborg in his *Principia Rerum Naturalium (Principles of Natural Things)* (Swedenbourg 1734). In this work, he theorizes that the planets were the result of a shell of material being expelled from the vicinity of a star. The shell initially rotates with the star. As the shell expands outward, it is stretched to the point that it breaks into fragments, at which time the underlying star is observable. The larger fragments become planets, and the smaller ones fall back toward the star. Unlike Isaac Newton, who published his *Principia* based on mathematics 47 years before (Newton 1687), Swedenbourg used a philosophic or rationalist approach in his theories that was largely devoid of mathematical proof (Baker 1983). At this time, Newton's Laws of motion and gravity were relatively new and still met with skepticism by some, particularly with respect to the true nature of the unseen gravitational force central to Newton's worldview. Indeed, it was another ~180 years before Einstein figured it out. Swedenbourg also published works on diverse topics such as anatomy, the functioning of the brain, and metallurgy. Later in life, Swedenbourg became a

DOI: 10.1201/9781315210643-5

spiritualist and published several works at the border between mysticism and religion, which led to the founding of "The New Church" after his death in 1772 (Swedenbourg 1907). The church continues to this day. Although Swedenbourg's approach to describing planet formation is not consistent with rigorous theory or observations, it is one of the first descriptions of a nebular theory for star and planet formation .

Immanuel Kant, well-known Prussian–German philosopher, was in the next generation after Swedenbourg and knew of his work on planet formation. By that time, telescope technology had improved to the point where gaseous nebulae could be observed in the vicinity of stars. Using the rationalist approach and knowledge of Newton's Laws, he theorized that stars form from the gravitational collapse of interstellar gas clouds. As the collapse proceeds, a gas cloud would spin up and, due to centrifugal force, form a disk which would later fragment into planets. He was on the right track but did not have the mathematical prowess to develop analytical models to test and push the limits of his intuition. He discusses his theory of solar system formation in the essay, "Universal Natural History and Theory of the Heavens" (Kant 1755; Johnston 2008). At the beginning of the same essay, he uses his powers of deduction and the appearance of the night sky to correctly infer that we live in a disk galaxy. Kant realized he could only take his theories so far, and at one point in the essay states, "But I prefer to leave this thought to those who find in themselves more reassurance in dealing with unprovable knowledge and more motivation to set down an answer". It would take the mathematical genius of the next generation researcher, Pierre-Simon Laplace (Figure 5.1), to bring our understanding of star and planet formation to the next level.

Pierre-Simon Laplace was born in the Normandy region of France in 1749. He was the son of a small landowner and cider merchant. His father sent him off to the University of Caen to study theology and become a priest. However, he fell in love with mathematics instead and left his religious studies before graduating. Armed with a letter of recommendation from his former math professor, Le Canu, he went to Paris in the hopes of studying with a well-known mathematician, Jean le Rond d'Alembert (O'Connor et. al. 2007; Whittaker 1949). Apparently, the first meeting between the two did not go well. To prove his worth, one account has it that d'Alembert gave Laplace a difficult problem to solve, which he did overnight. He was then given a harder problem, which he also solved overnight. d'Alembert was sufficiently impressed that he recommended Laplace for a teaching position at Ecole Militaire, a French military school in Paris. There he would later come to know a famous cadet, Napoleon Bonaparte. (Years after, Napoleon appointed, however briefly, Laplace to the post of Minister of the Interior (Gillispie and Grattan-Guinness 2000; Clerke 1911)). Now a professor, Laplace could devote his full energies to what he loved best, mathematics. During the next 17 years, he made many contributions to various fields of science, including mathematics, thermodynamics, and astronomy. The culmination of Laplace's work in astronomy is his *Mécanique céleste*, published in five volumes between 1798 and 1825. In it, Laplace summarized the history of astronomy and used the pioneering work of Newton as described in *Principia* as a foundation on which to develop mathematical descriptions that explain observed physical and astronomical phenomena from first principles without the need for empirical equations or Divine intervention (Woodward 1891). When Napoleon, who by that time was ruling France, asked

FIGURE 5.1 Pierre-Simon Laplace (1749–1827). French mathematician, also sometimes referred to as the "French Newton", who made many contributions to the physical sciences, including the Nebula Hypotheses for planet formation. Image Credit: Alamy.

Laplace how he had been able to write such a "large book" on how the universe works without mentioning its Creator, Laplace responded, "I had no need of that hypothesis". This amused Napoleon greatly (Ball 1908). The statement is noteworthy in that it is a radical departure from those made by many of his predecessors (e.g., Newton, Swedenbourg, and Kant). Laplace's work endures and is foundational to all physical sciences. Leafing through it, as is also the case with Newton's *Principia*, is a humbling experience.

In 1796 Laplace published a popular version of his work on celestial mechanics that provided an analytical description of the nebular hypothesis for planet formation proffered by Kant (Laplace 1796; Woolfson 1993). In his description, Laplace begins with an extended, approximately spherical, slowly rotating dust cloud (see Figure 5.2). As the cloud cools, its thermal pressure goes down, allowing its internal gravity to take over, resulting in collapse. As the cloud collapses, the conservation of angular momentum makes the cloud spin up and flatten along the rotation axis, thereby forming a disk (i.e., the pizza analogy of Section 4.5.2). In Laplace's model as the protostellar disk collapses the apparent outward centrifugal force felt by the fast-spinning material at the equator comes into equilibrium with the inward pull of gravity. As the collapse of the disk proceeds, the equatorial material in equilibrium with gravity is left behind, forming an annular ring in a circular Keplerian orbit about the central mass, i.e., the protostar (see Figure 5.2). As the disk continues to collapse, this process is repeated and multiple concentric

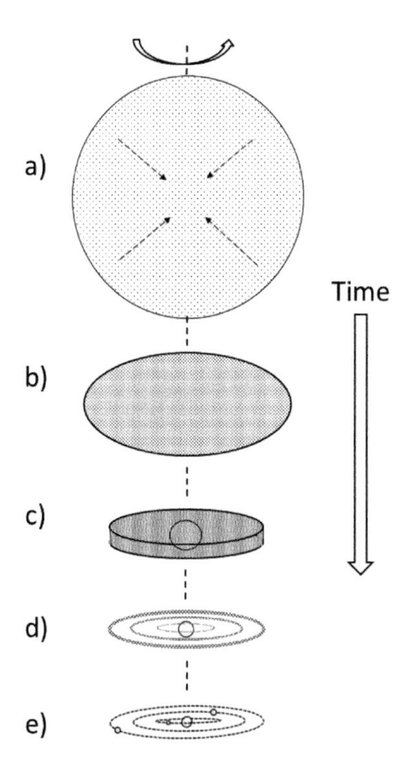

FIGURE 5.2 Laplace Nebula Hypothesis. In this model, the gravitational energy within a cooling, slowly rotating interstellar cloud overcomes its internal thermal energy and begins to collapse (a). Due to rotation, it preferentially collapses along its rotational axis, creating first a lenticular structure (b), and then a thin disk with a central protostar (c). Further contraction of the disk leads to the formation of concentric annuli (d), which eventually coalesce into planets moving in Keplerian orbits (e). The model is compelling but suffered from angular momentum issues, which were addressed by later investigators. Adapted from Woolfson 1993.

rings are formed. Small concentrations of mass (today called planetesimals) within each ring would ultimately coalesce to form a protoplanet. The protoplanet itself would undergo a similar collapse sequence leading to the formation of a ring system and, ultimately, moons.

This is a beautiful story based on Newtonian concepts, but it has a fatal flaw. Using his laws, Newton was able to estimate the relative mass and orbital velocities of the Sun and planets. With these estimates, it was possible to determine the distribution of angular momentum within the solar system. The Sun was found to contain 99.86% of the mass of the solar system, but <1% of its angular momentum, which is inconsistent with Laplace's collapse model. Once again, we find that even true geniuses like Aristotle, Laplace, Einstein, and others who push the boundaries of knowledge can sometimes get it wrong but in the process pave the way for future advancements.

Edouard Roche (1854) and James Jeans (1919) both tried to save Laplace's model. Roche invoked a preexisting condensed Sun, while Jeans suggested the initial mass of the nebulae was far greater than initially thought; neither of which seemed physical. Alternative theories for planet formation were investigated. These included a tidal model where the planetary material came from large solar prominences (Chamberlain

1901; Moulton 1905) or was pulled out of the Sun by a passing star (Jeans 1917). Later it was shown that tidal forces would only result in pulling material out to within a few radii of the Sun (Lyttleton 1960). Spitzer (1939) showed that even if solar material could be pulled out far enough, its thermal energy would be so high compared to its gravitational potential energy that the protoplanet could not hold itself together and would disperse into space (Woolfson 1993). Later a capture theory was proposed (Woolfson 1964; Dormand and Woolfson 1989), where planets formed as the result of tidal interaction between a newly formed Sun and a nearby collapsing protostar. By the early 1990s observers found that protoplanetary disks are common (Beckwith and Sargent 1993; Strom, Edwards, and Skrutskie 1993), making the likelihood of meeting the requirements of the capture theory in so many systems unlikely.

What is needed to make the solar nebular disk model (SNDM) work is a mechanism that transfers angular momentum radially outward from a much more massive protostar to the disk. This would be consistent with what is actually observed *and* permit the inward flow of material to the protostar through the accretion disk (Lüst 1952). The mixing of momentum provided by gas turbulence within the disk might be one such mechanism. There have been several theories for the origin of turbulence in disks, including thermal convection (Cameron 1978), magneto-hydrodynamic instabilities (Balbus and Hawley 1991), and purely hydrodynamic instabilities (Klahr and Bodenheimer 2004). In the turbulent model, the disk is assumed to have a viscosity (i.e., internal friction), ν, such that (Shakura and Sunyev 1973; Klahr and Bodenheimer 2003),

$$\nu = \alpha c_s H_D \tag{5.1}$$

where

ν = disk viscosity
α = dimensionless proportionality constant
c_s = sound speed in the gas
H_D = pressure scale height

The "α models" of disks have been very successful in helping to understand the formation and evolution of planetary systems. Assuming the disk is thin and in thermal equilibrium, equations describing the disk structure can be derived in terms of α. The value of α is a free parameter, but since many observables are only weakly dependent on its exact value, the theory remains applicable under a wide variety of assumed conditions. The Balbus and Hawley model (1991) utilizes the interaction between charged particles in a rotating disk and a weak axial magnetic field for angular momentum transport. A protostellar magnetic field could be the result of interstellar magnetic fields trapped in ionized gas participating in the collapse or created through a magnetic dynamo effect within the rotating disk (Blandford and Payne 1982) or some combination of the two. Magento-hydrodynamic (MHD) models of disks naturally produce the bipolar jets seen shooting out along the rotation axes of accreting protostars and black holes (see Figure 5.3). The complex, physical processes occurring in protoplanetary disks have pushed the limit of purely analytical analysis, such as that first done by Laplace over 200

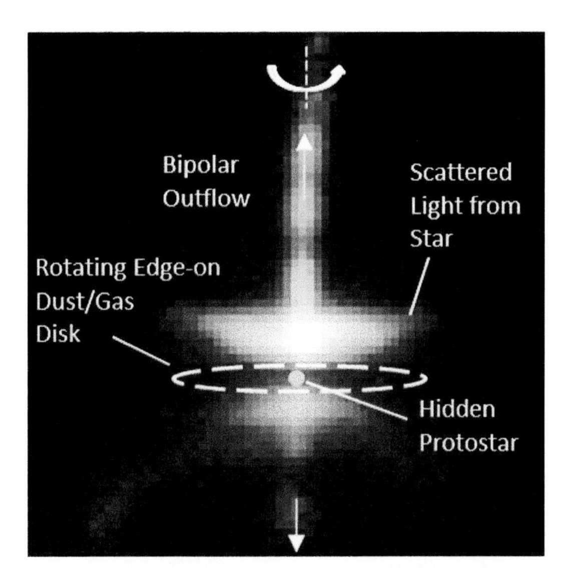

FIGURE 5.3 HH30. This nebulae Hubble Space Telescope image shows HH30 to be a protostellar system with an edge-on, 420 AU disk of dust and gas that obscures the central protostar. Scattered light from the protostar is seen directly above and below the disk. Jets of gas emanate along the rotation axis of the system, consistent with an MHD disk accretion model (Louvet *et. al.* 2018). Image Credit: C. Burrows (STScI & ESA), the WFPC 2 Investigation Definition Team, and NASA.

years ago. Today, the analytical approach to disk physics is augmented by numerical, three-dimensional, hydrodynamical simulations that run on computers (e.g., Boss 2017).

5.2 PROTOPLANETARY DISK OBSERVATIONS

Eighteenth- and nineteenth-century astronomers were unsure whether the spiral nebulae they observed with their telescopes were planetary systems in the process of formation or other galaxies. This uncertainty continued into the early twentieth century until the distance to the Great Spiral Nebula of Andromeda was measured. The distance was first estimated by Herber Curtis in 1917 using the brightness of a star in Andromeda going nova and then later by Hubble and Humason using the Cepheid variable period–luminosity relationship (see Section 3.5.1). The measured distance to Andromeda (~2.5 Mly), together with its angular width on the sky (~3°), indicated it was about the same size as the Milky Way (see Eq. 3.1). Its large size showed once and for all that Andromeda is not a young protoplanetary system but a galaxy of many billions of stars.

The closest protoplanetary disk surrounds the K7 star TW Hya at a distance of ~182 ly. The disk itself is relatively large (448 AU in diameter), but, due to its distance, only subtends ~8″ on the sky. One of the next closest disks is associated with the K5 star HL Tau and is just 2″ in diameter. In comparison, our eyes have an angular resolution of ~60″, making telescopes essential to the observation of protoplanetary disks. For ground-based observations, the situation is made more challenging by the presence of the Earth's atmosphere. When passing through the atmosphere, incoming photons can be absorbed or scattered by dust particles and/or air molecules. In terms of absorption, the worst offender is water vapor (e.g., clouds). This is particularly true at infrared and submillimeter wavelengths. At optical wavelengths, atmospheric turbulence can limit observations even on a clear night. Turbulent air motions can lead to time-varying pockets of high and low density, which causes the paths of photons to first bend one way and then another as they pass through them on the way to the observer. This is what causes the twinkling of starlight and leads to the blurring of images taken through a telescope. The magnitude of the bending is typically sufficient to wipe out images of arcsecond-sized protoplanetary disks. However, when the atmosphere is exceptionally calm, sub-arcsecond "seeing" with telescopes is possible. Under these rare conditions Herbig and Rao (1972) noted the nonpoint-like appearance of HL Tau, suggesting the presence of circumstellar material. A similar observation was made by Herbig (1968) toward the young, massive Ae/Be star R Mon.

The deleterious effects of the Earth's atmosphere on astronomical observations can be dealt with in several ways. Observatories can be located on high mountain tops, flown on aircraft, or suspended from high-altitude balloons; all with the goal of getting above as much of the absorbing, turbulent atmosphere as possible. In addition, adaptive optics can be added to telescope detection systems to straighten out the paths of incoming photons (Beckers 1993). This is most often accomplished by using a deformable mirror that can bend the paths of the incoming photons back to the way they would have been if the air molecules were not present. An alternative approach is to take a series of short-exposure images more rapidly than the atmosphere is changing and then shift and stack them in post-processing so that they line up. Once stacked, the images can be added together to yield a longer-exposure image with greater angular resolution than otherwise possible. This latter approach is called speckle imaging (Dyck and Howell 1982).

If possible, the best thing to do is to avoid the atmosphere altogether by putting the telescope in space. This is what was done to provide us Figure 5.3, which was taken at optical wavelengths toward the young K7 star, HH 30, in 1995 from an altitude of 547 km using the Hubble Space Telescope (HST). In this image, the 420 AU protoplanetary disk is seen edge-on. The dust in the disk blocks the light in the disk's equatorial plane. The dust thins out enough along the polar/rotation axis that scattered light from the hidden protostar can be seen in the form of bipolar cones. The dust grains can not only block or scatter the light but also absorb it as well. The grains are typically ~0.2 microns in diameter, about the same size as ultraviolet (uv) photons. This makes them effective absorbers of starlight. The absorption of the energy associated with uv photons makes the grains heat up, rendering them and their associated protoplanetary disks observable at infrared and far-infrared wavelengths.

Cohen (1983) was the first to combine optical, infrared, and far-infrared observations to characterize the protoplanetary disk around HL Tau. The far-infrared observations were performed at wavelengths of 52 and 100 microns using the Kuiper Airborne Observatory. The angular resolution was sufficient to suggest the presence of a nearly edge-on disk oriented in the northeast-to-southwest direction. A year later, further evidence for the disk was provided by Beckwith et al. (1984) using near-infrared speckle interferometry techniques to observe HL Tau with several ground-based telescopes.

These results motivated Beckwith et al. (1986) to attempt to observe the much more massive and extended gas component of the disk in the light of carbon monoxide (^{12}CO) molecules at

FIGURE 5.4 Owens Valley Radio Observatory (OVRO) interferometer. This interferometer, operated by Caltech, combined the light of an array of 10.4 m parabolic antennas to synthesize a millimeter-wave image comparable to what would be observed by a single telescope with a diameter equal to the maximum distance between antennas. The antennas were mounted on tracks so that the separation between them could be adjusted to provide the desired angular resolution. The greater the separation (i.e., baseline), the greater the effective angular resolution would be. The array was initially composed of 3 antennas and grew to 6. The University of California at Berkeley operated a similar array (i.e., BIMA) composed of nine, 6.5 m antennas. Both were used for the initial observations of protostellar disks at millimeter wavelengths. In Europe, the Plateau de Bure interferometer composed of six, 15 m antennas was also used for this purpose. Image Credit: Shutterstock.

millimeter wavelengths using the Caltech Owens Valley millimeter-wave interferometer (Figure 5.4). At the time, the interferometer was composed of three, 10.4-m dish antennas whose output was combined to create a beam on the sky whose size is inversely proportional to dish separation. By observing an object repeatedly with different dish separations and combining the results, it is possible to "fake-out" observing an object with a single telescope many meters or kilometers across. In the far-infrared and millimeter wavelengths such large telescopes are needed to provide the required angular resolution on the sky to observe protostellar disks. The initial observations with the Owens Valley interferometer detected ^{12}CO gas at the presumed location of the disk, but more sensitive, higher-resolution observations were required to characterize the disk itself. These observations were performed by Sargent et al. (1987, 1991) again using the Owens Valley interferometer, but this time tuned to the less abundant ^{13}CO molecule. A more abundant molecule like ^{12}CO may be brighter, but it also becomes optically thick easier, making it difficult or impossible to see deep enough into a protostellar cloud core to discern a disk. Sargent et al. did indeed see an edge-on protoplanetary disk, oriented northeast–southwest as suggested from the far-infrared dust observations by Cohen (1983). However, one great advantage of observing spectral lines is that their frequency can be shifted via the Doppler effect by the motions of the gas. Sargent et al. observed frequency shifts in ^{13}CO line emission across the protostellar disk consistent with the gas moving in Kepler orbits around the central protostar. These observations confirmed the presence of a protoplanetary disk around HL Tau (see Figure 5.5).

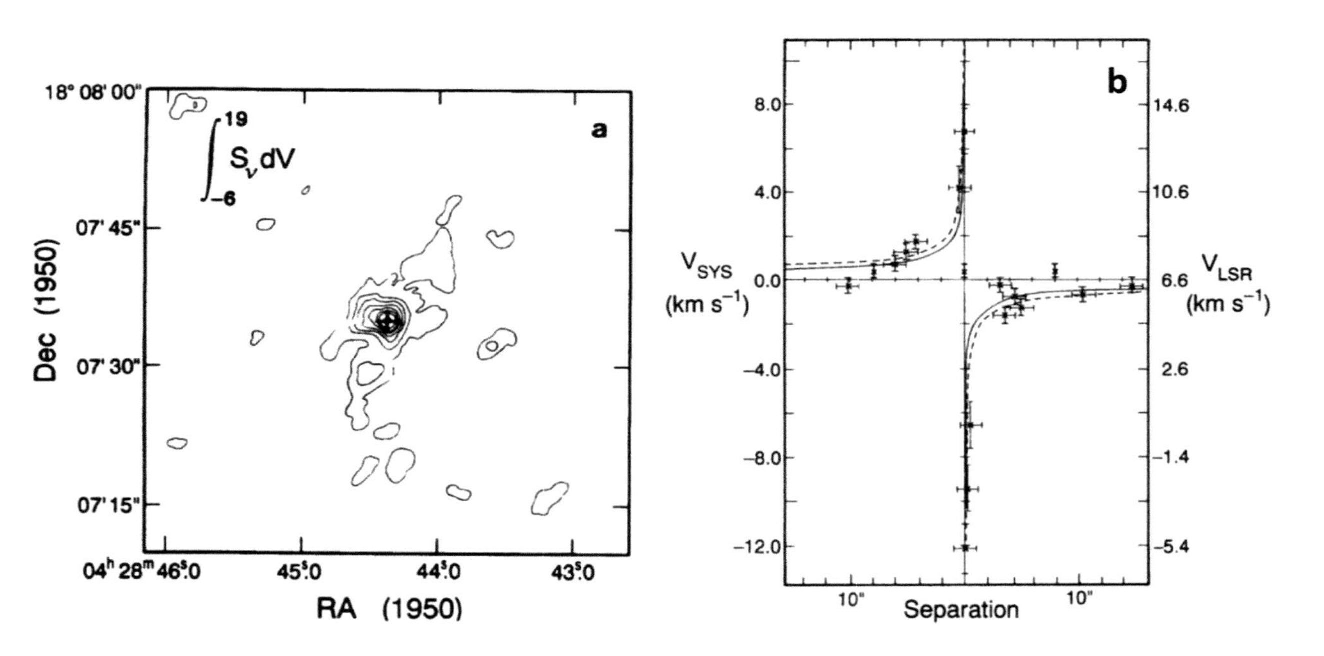

FIGURE 5.5 HL Tau Protoplanetary Disk. (a) A map of the ^{13}CO emission around HL Tau taken at an angular resolution of 2.7″ with the OVRO millimeter-wave interferometer (Sargent et al. 1991). The ^{13}CO emission was integrated over a –6 to 19 km/s velocity range. The disk is observed to be nearly edge-on with a northeast–southwest orientation. (b) The observed rotation curve for HL Tau derived from observed CO emission. The solid and dashed lines are what is expected for gas traveling in Keplerian orbits around a star of 0.55 and 1 solar mass, respectively. The fit between theory and observation is consistent with the presence of a protoplanetary disk.

Contemporaneous with some of the observations of HL Tau, the Infrared Astronomical Satellite (IRAS) conducted the first all-sky survey at far-infrared wavelengths (Neugebauer et al. 1984). One of the brightest objects revealed by the survey was IRAS16293-2422. The object is so deeply buried in the Rho Ophiuchi molecular cloud that its presence is only revealed at far-infrared wavelengths. Subsequent dust continuum and molecular line observations with the National Radio Astronomy Observatory 12 m telescope on Kitt Peak, Arizona, showed it to be a collapsing protostellar object with an extended, rotating disk (Walker et al. 1985, 1986). The presence of two bipolar outflows from the disk suggested the object is a binary system containing two protostars each having its own accretion/protoplanetary disk. High angular resolution observations with the Berkeley Illinois Maryland Array (BIMA) confirmed the binary nature of the system (see Figure 5.6; Walker et al. 1993).

In a quest for ever greater angular resolution and sensitivity, over the next 20 years, a consortium of countries, including Europe, the United States, Canada, Japan, South Korea, Taiwan, and Chile worked together to design and build the Atacama Large Millimeter Array (ALMA). Located at an altitude of 5,059 m (16,597 ft) in the dry desert of northern Chile, ALMA is an interferometer consisting of 66 antennas that, when fully extended (i.e., maximum baseline), has the angular resolution of a telescope 16 km in diameter. ALMA can perform observations over a wavelength range from 3.6 to 0.32 mm (31–1,000 GHz). In comparison, the largest single-aperture telescope (i.e., FAST in China) is "just" 0.5 km in diameter and can only operate up to 3 GHz. At its highest frequency of operation and longest baseline configuration, ALMA can achieve an angular resolution of ~0.005″. With this angular resolution it is possible to resolve protoplanetary disks in nearby star-forming regions, such as the one containing HL Tau, to size scales <1 AU. An ALMA image of the thermal (i.e., blackbody) emission from dust associated with the HL Tau protoplanetary disk is shown in Figure 5.7. The angular resolution is sufficient to resolve the 235 AU disk into concentric dust/gas rings. The appearance of this system is not so different from what was imagined by Swedenbourg, Kant, and Laplace over two centuries before.

5.3 FORMATION OF PLANETARY SYSTEMS

Substructures (i.e., rings, gaps, spirals) have been found to be common features in protoplanetary disks (see Figure 5.8). Such features can also be seen in the ring system of Saturn (e.g., the Cassini Division, see Figure 5.9). As described by Andrews (2021), theories for the origin of disk substructures generally fall into two broad categories. Most likely, the correct answer will end up being a combination of the two. The first is that the disk substructures arise from turbulent and/or hydrodynamic instabilities in the disk that lead to pressure traps that tend to concentrate solids into rings. The greater the density of solids, the greater the chance that collisions will occur, leading to the formation of planetesimals within the rings. Planetesimals are small rocky and/or icy bodies that through collisions and adherence to one another could over the course of time build up into a planetary body. This is the essence of the planetesimal theory of planet formation. Comets and asteroids are examples of left-over planetesimals from the formation of the solar system. The second

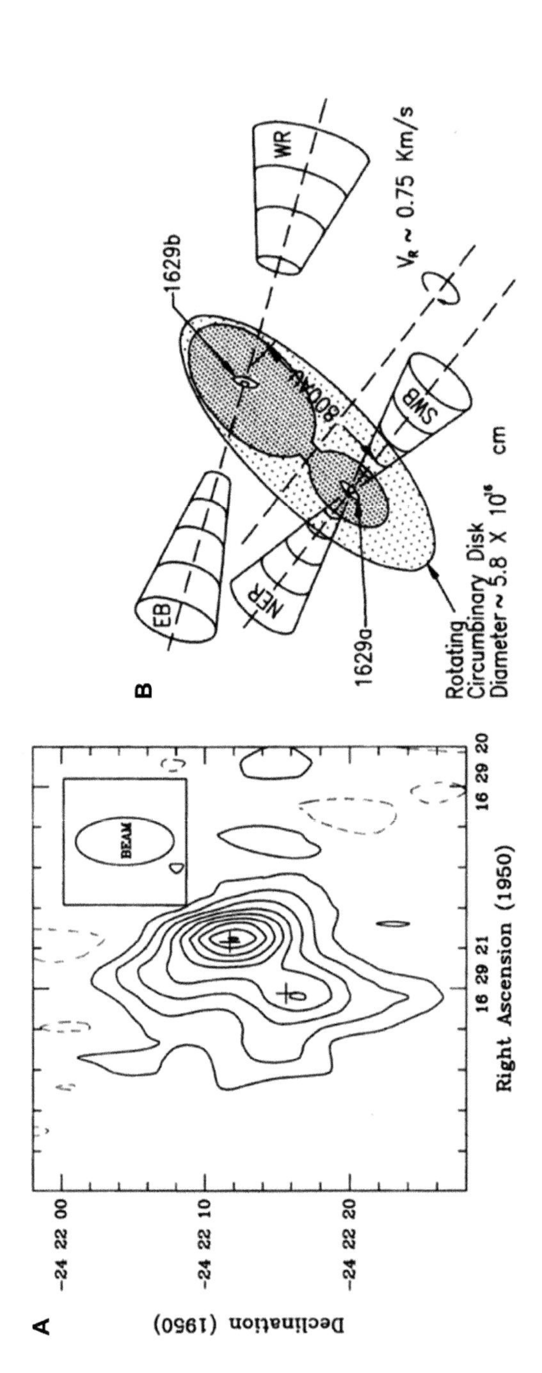

FIGURE 5.6 IRAS16293 Protobinary System. (a) Interferometric map of the 3 mm continuum emission. (b) Schematic of the system. Two protostellar objects, 16291 and 1629b, are embedded in a rotating circumbinary disk of material. The outflow system driven by 1629a is defined by lobes NER and SWB. 1629b is shown as the driving source for the outflow defined by lobes EB and WR (Walker et al. 1993).

FIGURE 5.7 ALMA image of HL Tau Protoplanetary Disk. The image was taken using 25–30 antennas with a maximum baseline of 15.24 km at 233 GHz. The observation required 4.5 hours of integration time. The resulting angular resolution was 0.035″ or about 5 AU at the distance of HL Tau. The 235 AU disk contains concentric bright rings that may indicate the presents of recently formed protoplanets in the gaps in between. Image Credit: ALMA (ESO/NAOJ/NRAO).

category of models are based on the proposition that disk substructures are generated by interactions between existing protoplanets and the disk. As a young planet moves in its orbit about a star, it will sweep up material (e.g., planetesimals), initially through collisions and later through gravitational attraction, leaving a gap behind. (The process is analogous to rolling a snowball in a yard to make the larger balls for the torso of a snowman. If the yard is thinly covered, once the ball accumulates enough mass, it will start accreting all the snow underneath, leaving a bare surface in its wake.) If there is sufficient material along the planet's orbital path to allow it to accrete a few Earth masses, the planet will have sufficient gravity to rapidly acquire an atmosphere tens or hundreds of times greater than the mass of the Earth and become a gas giant, like Jupiter and Saturn. This theory is generally referred to as the core-accretion model of planet formation.

Planetesimals containing significant amounts of water ice are believed to provide both the additional mass and "stickiness" needed for the formation of the "super-Earth" cores of Jovian planets. Such icy planetesimals can only exist at distances sufficiently far from their parent star that their constituent ice does not sublimate away as a result of the star's radiated heat. The orbital distance at which this condition is met is referred to as "the snowline" or sometimes "the frostline" (see Figure 5.10). Its location is a function of the parent star's

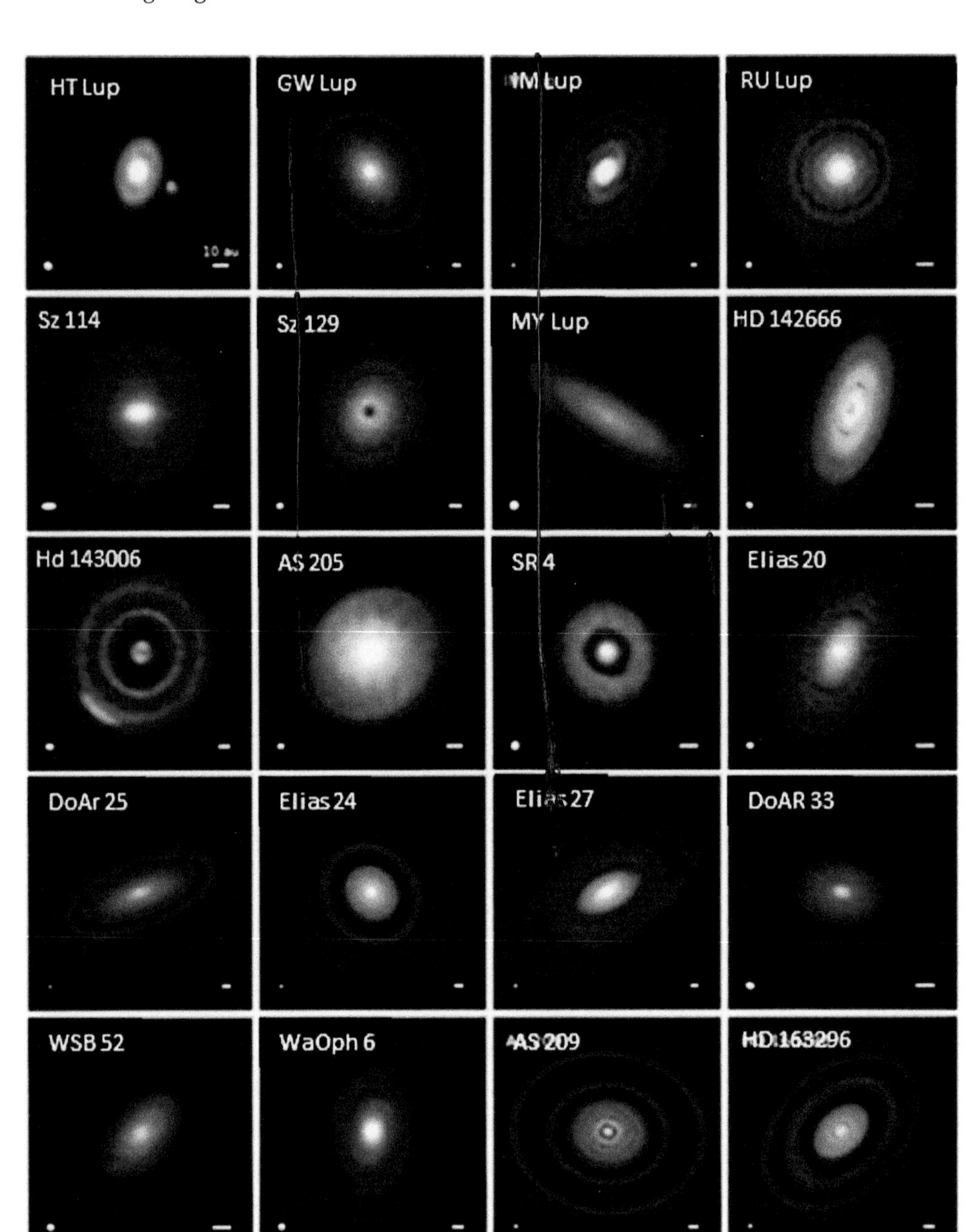

FIGURE 5.8 Gallery of protoplanetary disks. These images of the dust disks associated with forming planetary systems were made with ALMA at a wavelength of 1.25 mm (240 GHz). Representative beam sizes and 10 AU sidebars are provided in the bottom left and right corners of each disk image. Credit: Andrews et al. 2018.

luminosity and the disk's opacity. In the case of our own solar system, the snowline is currently located at an orbital radius of 5 AU. During the formation of the solar system, the Sun was less luminous than it is now and the amount of material in the ecliptic plane (the imaginary plane in which the planets orbit the Sun) greater, leading to lower temperatures. The lower temperatures resulted in the snowline being just ~2.7 AU from the Sun in what

FIGURE 5.9 Saturn Ring System imaged with the Cassini spacecraft. The rings of Saturn have features similar to those observed in protoplanetary disks. An example of which is the Cassini Division, a large gap in the rings discovered in 1675 by Giovanni Cassini using the Paris Observatory refracting telescope. The Moon Mimas is seen in the background. Credit: NASA/JPL/Space Science Institute.

is now the asteroid belt between Mars and Jupiter. This theory of planet formation explains why the gas giant planets (Jupiter, Saturn, Uranus, and Neptune) are found in the outer solar system, while smaller rocky, terrestrial planets are found in the inner solar system. In the case of HL Tau, all but a small region within the inner dark ring of Figure 5.8 is outside the snowline, suggesting that the dark rings in the image could be associated with the formation of multiple gas giant planets. Gravitational interactions between planets or between a planet and disk can rob a planet of orbital angular momentum, causing it to spiral inward toward the central star. Such planetary migration appears to be common and explains why gas giants are sometimes found within the snowline of some planetary systems (see Section 5.4). Observations and theoretical models suggest that planets begin to form while the parent star is still in the process of accreting matter through its disk. Amazingly, the time it took for our Sun and its entourage of planets to form from a collapsing, light-year-sized interstellar gas cloud was only ~10 million years – just the wink of an eye in cosmic time.

Today left-over planetesimals from the formation of our Sun can be found scattered across the solar system. These include rocky asteroids which are found within the

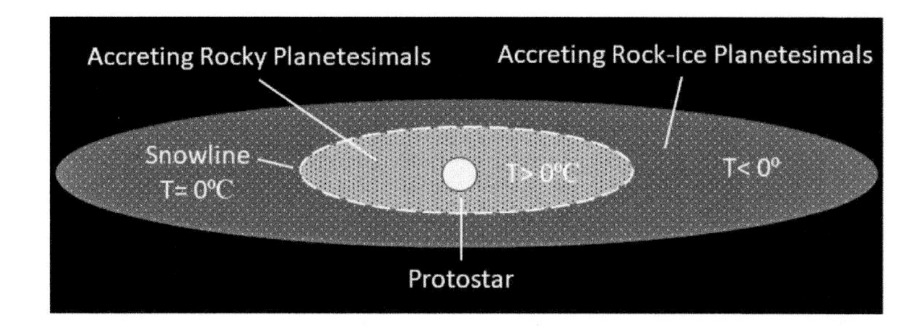

FIGURE 5.10 Snowline. The Snowline marks the distance from the protostar where temperatures within the protoplanetary disk drop below the freezing point of water. Beyond this radius planetesimals can contain both rock and ice. Ice can both "bulk-up" a planetesimal and makes it "stick" better to other planetesimals it collides with, thereby facilitating the creation of larger planetary bodies.

snowline and cometary nuclei composed of rock and ice that can be found primarily outside the snowline. Many of the original planetesimals suffered orbital angular momentum loss through disk interactions or drag from collisions and ended up spiraling into the Sun. Many others in the outer solar system were hurled into deep space by gravitational interactions with Jupiter and/or Saturn. Millions of such bodies did not fully escape the Sun's gravitational pull and are even now orbiting us at a distance of ~1 light year. This assembly of outcasts was first hypothesized to exist by Estonian astronomer Ernst Opik (1932) and later, independently, by Jaan Oort (1950). Today the assembly is referred to as the Opik–Oort Cloud (see Figure 5.11). Occasionally interactions with a passing star can nudge one of the wayward cometary nuclei onto a trajectory to the inner solar system. Comets lose prodigious amounts of gas and dust

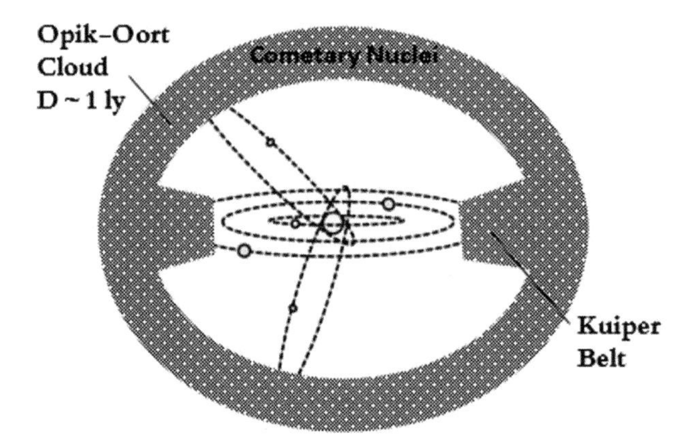

FIGURE 5.11 Schematic of Opik–Oort cloud. Located ~1 ly from the Sun, the Opik–Oort cloud is a spherical cloud containing millions of cometary nuclei that are left-over planetesimals from the formation of the solar system. Occasionally gravitational interactions cause one or more of them to pass through the inner solar system. The Kuiper Belt is a remnant of our protoplanetary disk that extends between 30 and 50 AU from the Sun. It contains many small bodies, including cometary nuclei and dwarf planets.

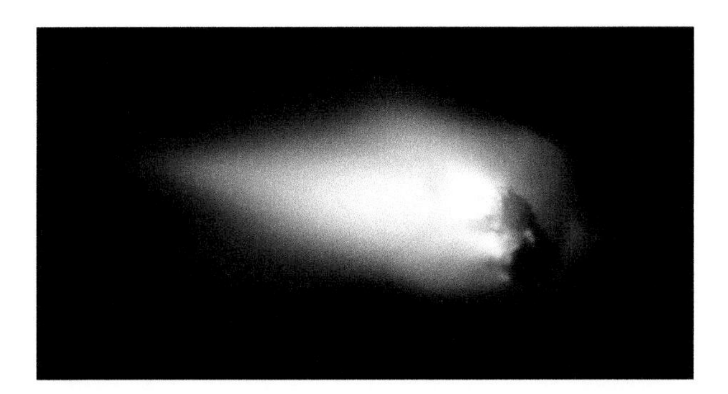

FIGURE 5.12 Halley's comet. This image was taken by the European Giotto spacecraft on March 13, 1986, as it flew past Halley's comet. The comet nucleus is clearly visible on the left and is 9.3 × 5 miles (15 × 8 km) in size and rotates on its axis every 7.4 days. It is a loose compaction of rock and ice, forming a "rubble pile". The comet is believed to have originated in the Oort cloud. However, gravitational interactions with the outer planets have placed it in a highly elliptical orbit that brings it near the Sun every 75–76 years. When this occurs heat from the Sun sublimates ice on or near the comet's surface leading to the formation of a coma, a nebulous gas/dust cloud around the nucleus, and an extended tail. Credit: Halley color Camera Team, Giotto Project, ESA.

if they pass close enough to the Sun for their ice to start melting. The melting releases a cloud of gas and dust around the comet's nucleus producing a cometary coma and a tail of debris extending for millions of kilometers back along its trajectory (Figures 5.12 and 5.13). By studying this material with Earth and space-based telescopes, we can gain insights into the conditions within our primordial protoplanetary disk. In recent years, spacecraft have visited both a comet (Figure 5.14; Rosetta 2015) and an asteroid (Figure 5.15; Lauretta et al. 2019), allowing detailed studies of their composition and structure. Glimpses into our solar system's past can also be got through studies of Kuiper Belt Objects. The Kuiper Belt is a fossil remnant of our extended protoplanetary disk that contains many small bodies, including dwarf planets like Pluto.

5.4 FORMATION OF TERRESTRIAL PLANETS

The accumulation of evidence for the formation of terrestrial planets from the collision and accumulation of planetesimals began with the first telescopic observations of the Moon by Galileo. When a new astronomical instrument is developed, often the first thing a developer does to validate its performance is to look at a bright, well-known object. This is what happens today and was also true in Galileo's time as well. Between November 30 and December 18, 1609, Galileo conducted an observational study of the Moon using a telescope of his own construction with a 20× magnification. What he found shattered the long-held Aristotelian belief that the Moon is a perfectly smooth sphere. Galileo noted that the boundary (i.e., the terminator) between the day (sunlit) and night (shadow) parts of the Moon was not a straight line as would be the case if the Moon was smooth, but irregular (see Figure 5.16; Galileo 1610). He also noted that the appearance of the terminator changed over time. He correctly deduced that these features were caused by the movement of shadows cast by elevated surface features, such as mountains and ridges around craters.

FIGURE 5.13 Harbinger of Fortune. Halley's comet becomes bright enough to be seen with the naked eye. Its appearance was recorded in 240 BC by Chinese astronomers and perhaps earlier by the Greeks (Kronk 1999). Its appearance in 1066 AD is recorded in the Bayeux Tapestry by medieval Europeans. The inscription to the right of the comet says "They marvel at the star." Edmond Halley (friend and publisher of Isaac Newton) was the first to compute the comet's orbit and showed the same comet had been observed multiple times in the past. Image: Shutterstock.

Indeed, every terrestrial surface that has been imaged (albeit a planet, Moon, asteroid, or comet) shows impact cratering from an onslaught of smaller bodies. The older the surface, the longer the time has been to accumulate impacts. Therefore, the more craters that are observed, the older the surface is estimated to be (see Figure 5.17). If the planetesimal theory of planet formation is correct, one might then expect to see all planetary surfaces covered with craters associated with the last wave of planetesimal impacts from their formation. However, old impact craters can be "washed away" by erosion (e.g., weathering, volcanism, or plate tectonic processes on Earth) or filled in by molten magma from subsequent impacts (e.g., Meteor Crater, Figure 5.18). Indeed, the best-preserved impact craters on Earth are found in deserts (see Figure 5.18).

From isotopic dating of Moon rocks brought back by Apollo astronauts, the last major wave of planetesimal impacts is estimated to have occurred between 4.1 and 3.8 billion years ago, a time referred to as the Late Heavy Bombardment (LHB). One theory is that the LHB was triggered by an orbital migration of Jupiter and Saturn that produced gravitational instabilities that scattered billions of planetesimals in what remained of our protoplanetary disk. Swarms of these objects found their way to planetary bodies in the inner

FIGURE 5.14 Left-over planetesimal from the outer Solar System. (a) Comet 67P/Churyumov–Gerasimenko as observed by the *Rosetta* spacecraft in 2014 photo: [ESA/Rosetta/NAVCAM]. The 4.3 by 4.1 km (2.7 by 2.5 mi) body originated in the Kuiper Belt. b) A picture taken by the *Philae* lander on November 12, 2014, from the comet's surface. Compared to where we were in the previous figure, our understanding of the physical universe has come a long way in a ~1,000 years! What revelations will the next millennia bring? Image Credit: ESA.

FIGURE 5.15 Left-over planetesimal from Inner Solar System. a) Asteroid 101955 Bennu as observed by the *OSIRIS-REx* spacecraft in 2019 [NASA photo]. The 0.5 km sized body is a carbonaceous asteroid in the Apollo group. (b) On October 2020, *OSIRIS-REx* touched down on the surface of Bennu and collected a sample using an extendable arm with a vacuum cleaner [Credit: NASA photo]. The sample will be returned to Earth in 2023.

and outer solar systems, the impacts of which led to the beleaguered appearance of many planetary surfaces (Gomes et al. 2005).

5.5 FORMATION OF THE EARTH

The only world in which we know life has evolved is our own. Life on Earth appears to have originated either in surface oceans, shallow seas, or tidal pools (see Chapter 6). Therefore,

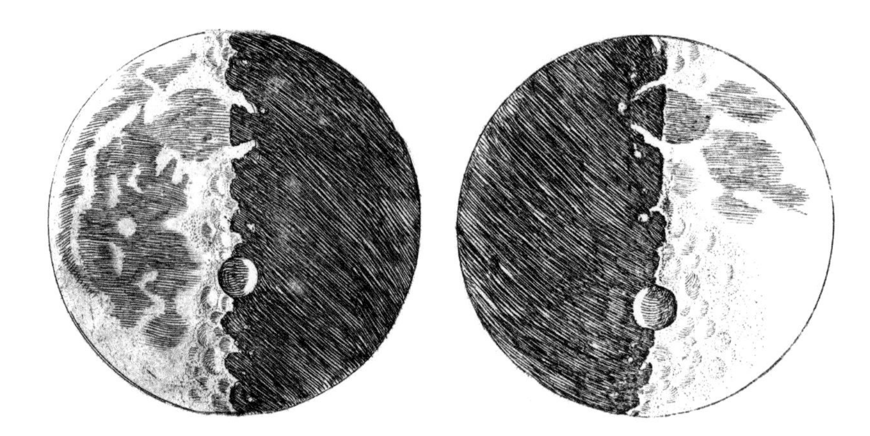

FIGURE 5.16 Galileo's sketch of the Moon. From observations made on the evening of November 30, 1609, using a telescope with 20× magnification, Galileo correctly concluded that the irregular appearance of the lunar terminator was due to mountains and ridges casting shadows. The first sketch of the Moon made using a telescope was drawn by the English astronomer Thomas Harriott on July 26, 1609. However, his sketches were inaccurate and his Dutch built, 6× telescope prevented him from seeing the level of detail provided by Galileo's more powerful, homemade telescope. Galileo published his results in 1610, while Harriott's were not printed until 1784. (Alexander 1998; Massey 2009). Image Credit: Alamy.

FIGURE 5.17 Callisto. Discovered by Galileo a month after he made the sketch of the Moon (Figure 5.16), Callisto is the second largest Moon of Jupiter. Its heavily cratered surface free of erosion and the influence of geological processes makes it the oldest in the solar system. Image Credit: NASA.

FIGURE 5.18 Meteor Crater. Formed by the impact of a 50 m iron–nickel meteorite ~50,000 years ago, its age and location in the Arizona desert make it the most pristine meteor crater of its size on Earth. Using his telescope, Galileo could have easily resolved a similar-sized crater on the Moon. Image Credit: NASA.

examination of how terrestrial planets like Earth formed can provide insights into identifying suitable worlds where life may have evolved elsewhere. The Earth began in a maelstrom of planetesimal collisions within the Sun's protoplanetary disk. It is one of a handful of surviving planetary siblings (the others being Mercury, Venus, and Mars) that formed within the disk's snowline (see Figure 5.10). The other siblings were either destroyed through collisions, consumed by their brethren, ended their existence by spiraling into the Sun, or were hurled into deep space through tidal interactions. Indeed, the elliptical orbit the Earth now finds itself in is the result of tidal interactions between the Sun, Jupiter, and Saturn.

Approximately 4.54 billion years ago the Earth was a ball of molten magma heated through the transformation of orbital kinetic energy into thermal energy by the planetesimal collisions that formed it (see Figure 5.19). The planetesimals contained all the raw materials needed to build a terrestrial planet, including iron (31.2%), oxygen-bearing compounds (30.1%), magnesium (13.9%), sulfur (2.9%), nickel (1.8%), calcium (1.5%), aluminum (1.4%), and trace elements (1.2%). Since the Earth was molten, the pull of gravity made the heavier elements sink toward its center. In geology this process is referred to as differentiation. Over millions of years, this resulted in the organization of the Earth's interior into chemically and structurally distinct layers (see Figure 5.20).

5.5.1 Formation of the Moon

One glaring piece of evidence for planetary cannibalism in the early history of the solar system can be seen staring down at us most nights, the Moon. Just as the internal structure

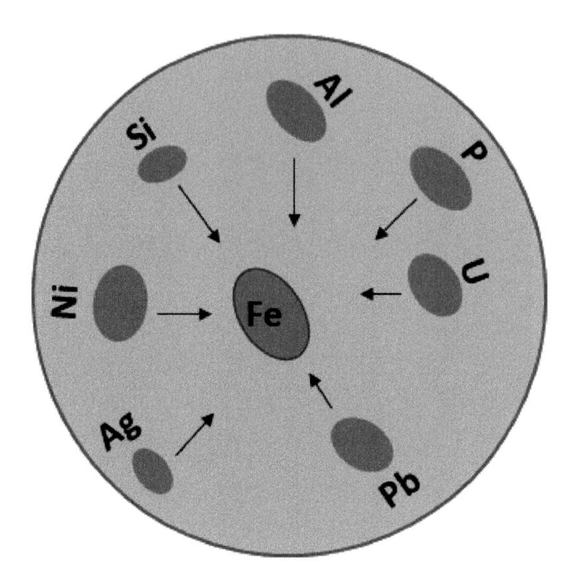

FIGURE 5.19 Proto-Earth. After forming from planetesimal collisions, the Earth was a sphere of molten lava with a thin crust. The heavier elements preferentially sunk toward the Earth's center through the geological process of differentiation.

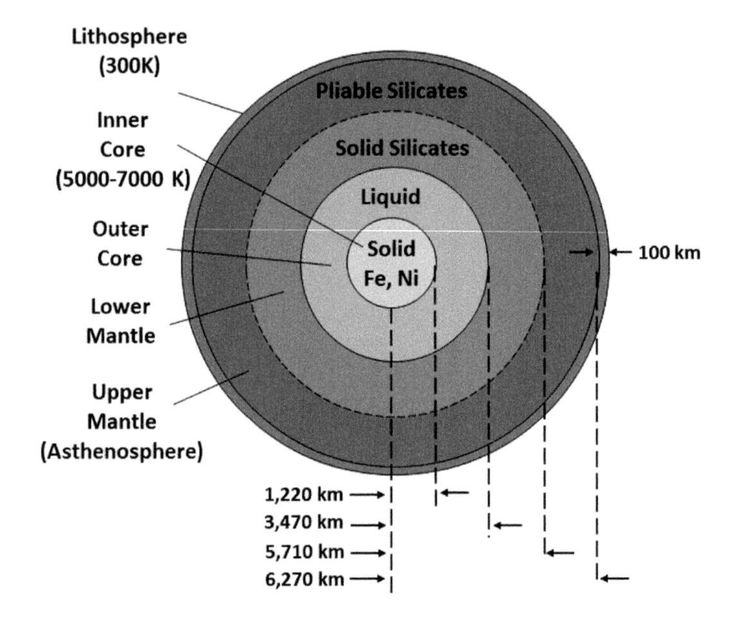

FIGURE 5.20 Earth's internal structure. Over millions of years the differentiation process resulted in density, temperature, and chemical gradients in the Earth's interior. Together with pressure, it is the combination of these conditions that determines whether material at a given depth will exist in a solid or liquid state.

of the Earth was beginning to stabilize, it was nearly destroyed by the impact of a planetary body. According to the Giant Impact Model for the origin of the Moon, a planetary body, called Theia, about half the size and one-tenth the mass of Earth shared an orbit sufficiently close to our own that gravitational interactions eventually put it on a collision course with the Earth (Figure 5.21). Most computer simulations (and the fact that

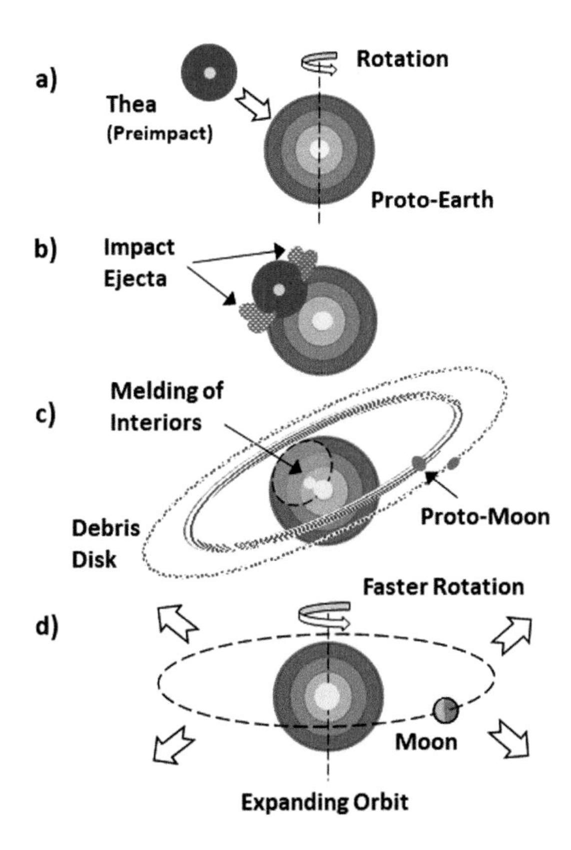

FIGURE 5.21 Origin of the Moon. The Giant Impact Theory is the prevailing model for the origin of the Moon. (a) A Mars size body collides with Earth ~95 million years after formation. (b) The collision is off-center and sends ejecta from the colliding body and the Earth's mantle into orbit. (c) Most of the colliding bodies' mass melds with the Earth's interior. (d) Over a few million years material in the debris disk coalesce into a proto-Moon that sweeps up any remaining ejecta. Due to the addition of angular momentum from the collision, the Earth rotates faster on its axis (once every 5 hours) than it did before. Tidal interaction with the Moon slows the Earth's rotation and accelerate the Moon's orbital velocity causing its orbit to expand over time. This process continues today.

the Earth remained largely intact) suggest it was not a direct hit, but a glancing blow. The impact ejected material into an orbit about the Earth. What iron-nickel core Theia may have had merged with that of the proto-Earth. Within a few million years, multiple satellites may have formed from the orbital debris, with the proto-Moon being the most massive. Computer simulations and elemental comparisons between the Moon and Earth suggest the impact occurred ~95 million years after the initial formation of the proto-Earth (Jacobson 2014). The evidence for the Giant Impact Model includes the similarity between the isotopic composition of minerals in the Earth's crust with the Moon rocks returned by the Apollo astronauts (Paniello & Moynier 2012; Saal et al. 2013). Other theories for the origin of the Moon include it forming from a common planetary disk as the Earth or that it was captured. Neither theory is consistent with the observed high angular momentum of the Earth–Moon system (Stevenson 1987).

The giant impact imparted not only additional mass but also additional angular momentum. After the collision the Earth rotated on its axis once every 5 hours (!) and the Moon rotated on its axis as well. The surface of the Moon was a magma ocean and would have glowed a dull red color. After its formation the Moon's orbital distance was between just 15,000 and 20,000 miles, making it appear ~15× bigger in the sky than it does today (Powell 2018). Over time tidal interactions slowly converted some of the Earth's rotational energy into lunar orbital acceleration, with the effect that the Earth's rotation slowed down and the Moon's distance increased to today's values. This process continues even now with the Moon's distance increasing by 1.5 in (38 mm) per year (Williams and Boggs 2016) and the Earth-day lengthening by 17 microseconds per year (Murray and Dermott 1999). One difficulty with the Giant Impact Theory is that one would expect a significant part of the Earth's surface would have been remelted due to the impact. However, there is no clear evidence that surface material was reprocessed by a magma ocean (Jones 1998) (Figure 5.22).

5.5.2 Uncovering the Inner Workings of the Earth

The internal structure of the Earth was determined by monitoring the propagation of seismic waves generated by earthquakes (see Figure 5.23). Earthquakes originate in the Earth's crust, typically along a boundary (i.e., fault line) between the plates (see discussion below). The location on the surface just above the location of the earthquake's point of origin is

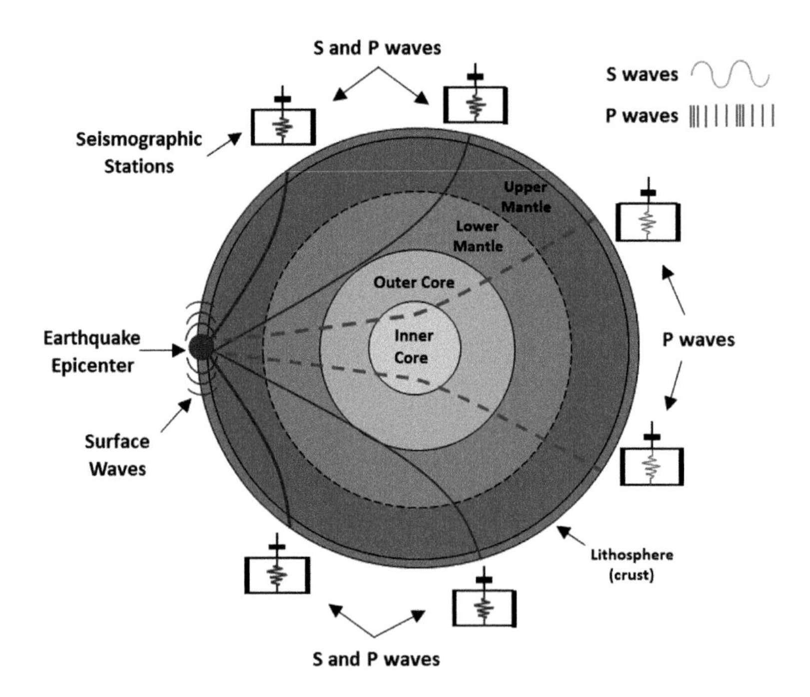

FIGURE 5.22 Utility of Seismic waves. Seismic waves are of three types: Surface waves that are restricted to travel on the surface; S-waves that can travel through solids; and P-waves that can travel through solids and liquids. By monitoring the arrival times of the waves at seismic stations around the globe the density, boundaries, and phases (i.e., whether solid or liquid) of the various layers of the Earth can be determined.

FIGURE 5.23 First Seismometer. The seismometer was invented by the Chinese astronomer Zhang Heng and used to detect a distant earthquake in 132 AD. The original version was 2 meter diameter bronze vessel with a pendulum in the middle. When the ground shook the pendulum would move, tripping a lever that released a ball that dropped into a toad's mouth. The position of the toad indicated the direction of the earthquake. Credit: Shutterstock.

called the epicenter. There are three types of seismic waves: surface waves, S-waves, and P-waves. Surface waves (sometimes called land waves) propagate along the surface and can be either transverse (i.e., up and down, in which case they are classified as a Rayleigh wave) or shear (i.e., moving from side to side, referred to as a long wave). S-wave and P-waves travel through the Earth's interior. S-waves are also transverse waves with up-and-down motions transverse to their direction of propagation. P-waves are pressure waves. They, like sound waves, characteristically create regions of high and low pressure along their direction of propagation. S-waves can travel through solids, but their back-and-forth (sloshing) motion is readily attenuated in liquids. P-waves, on the other hand, can travel through both solids and liquids (e.g., you can hear things clearly under water). Therefore, by placing seismic monitoring stations around the globe and monitoring the strength and arrival times of seismic waves, it is possible to determine the structure and phases (i.e., solid or liquid) of material throughout the Earth's interior without picking up a shovel. The seismometer was invented by Chinese astronomer Zhang Heng in the second century and used to notify the emperor if a major earthquake occurred (see Figure 5.23).

5.5.3 The Active Earth

The densest part of the Earth is the inner core, consisting of iron and nickel. The core of the Earth is estimated to host temperatures of between 5,000 and 7,000 K. These high temperatures are a result of the residual heat from the formation of the Earth and the decay of radioactive elements (e.g., uranium-238 and thorium-232) and is exacerbated by the crushing gravitational pressure of overlying layers. Indeed, if the overlying layers were somehow removed, the inner core of the Earth would glow yellow, like a small Sun. Enveloping the

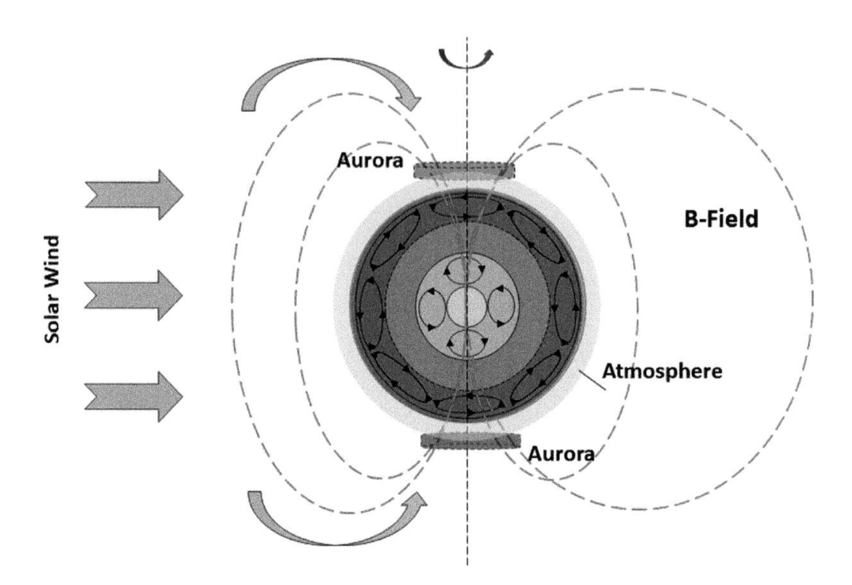

FIGURE 5.24 Convection cells. A combination of temperature, pressure, and composition result in two concentric layers within the Earth being pliable enough for convection cells to be established. The deepest is in the outer core, where the convective motion of liquid iron and nickel acts as current loops that generate the Earth's magnetic field. The second is in the aesthenosphere, located just under the Earth's crust.

inner core is a 1,400-mile (2,260 km) deep sea of molten iron and nickel, referred to as the outer core. If it could, the inner core would be liquid as well, but the pressure of the overlying layers keeps it in the solid phase. One can think of the outer core being analogous to a pot of liquid sitting on the hot burner of a stove. The liquid on the bottom absorbs heat from the burner. The absorbed heat increases the kinetic energy of the atoms or molecules making up the liquid, causing them to bounce around and increasing their relative separation forming low-density blobs that can float up. As the blobs float up, they transfer their excess heat to the surrounding material until they become as dense as the surrounding material, at which point they no longer float. If at the top of their motion they are further cooled by the overlying layer, their density will become greater than the liquid below, and they will sink. Once they reach the bottom of the pot, they will heat up again and the process will repeat, thereby forming a convective cell. This situation is depicted in Figure 5.24. Here we see that the heat of the inner core is sufficient to establish convection cells within the outer core. Since the hot iron and nickel are partially ionized, the convection cells form electric current loops. As demonstrated by Michael Faraday in 1831, current loops generate magnetic fields. Similarly, the current flowing within the convective cells of the outer core of the Earth is responsible for the generation of the Earth's magnetic field. It is amazing to think that the motion of a compass needle on the Earth's surface is directly tied to the motions of material ~2,800 km below.

5.5.4 Earth's Force Field

The Earth's magnetic field extends outward between 6 and 10 Earth radii, creating a magnetosphere around the Earth (see Figure 5.24). To charged particles magnetic fields appear as "rubber bands" that resist their passage. In electromagnetic theory this behavior is described by the Lorentz Force Law. Lucky for us, as a result of the Lorentz Force, the magnetosphere of the Earth acts as an invisible force field protecting us from the solar wind. The solar wind is a stream of charged particles consisting of electrons, protons, and alpha particles ejected into space from the Sun as a result of coronal and surface processes (e.g., solar flares). Even so, during solar outbursts, the solar wind can damage satellites and reach the Earth's surface, interrupting communication systems and causing power outages. During major solar outbursts the wind can pose a serious threat to astronauts. The solar wind compresses the magnetosphere on the sunward side and drags it out into a tail on the leeward side. (Think of an incoming wave enveloping your feet on the beach.) Some of the charged particles get trapped and spiral around the Earth's magnetic field lines until they plummet into the atmosphere at the north and south magnetic poles. It is the interaction of these incoming particles with atmospheric oxygen and nitrogen atoms that produce the colors of the northern and southern lights. The inner boundary of the magnetosphere is the ionosphere. The ionosphere is composed of layers of charged particles extending from 30 to 600 miles above the Earth's surface. The charged particles arise principally from ultraviolet (uv) light from the Sun ionizing atoms and molecules in the upper atmosphere. These charged particle layers mark the outer boundary of atmospheric electrical activity. The layers can also act as mirrors to radio waves, making over-the-horizon communications without the aid of satellites possible.

5.5.5 A Game of Continental Billiards

Surrounding the outer core is the lower mantle. The lower mantle, also known as the mesosphere, is a ~1,400 mile thick layer of rigid silicates (e.g., silicate perovskite) and ferropericlase; minerals that form under pressure at great depths. The upper mantle, also known as the asthenosphere, extends from just below the Earth's crust, i.e., lithosphere, to a depth of ~430 miles. The asthenosphere is also composed of silicates which become pliable at temperatures (~1,600 K) and pressures occurring just below its boundary to the lithosphere. Similar to the situation in the outer core, the pliability of the layer together with the presence of a hotter mesosphere underneath leads to the establishment of convective cells (see Figure 5.25). The speed of motion at the top of the convective cells just below the lithosphere is ~1 cm/year (about as fast as finger nails grow). The Earth's crust, or lithosphere, is composed of lower-density silicates and floats on top of the asthenosphere, like wood on water. The thicker part of the crust (~70 km) makes up the continental plates. They are composed of less dense silicates, e.g., granite. The lithosphere is only 5 and 10 km thick in the ocean basins, where it is composed primarily of basalt. It is here, where the crust is thinnest, that upwelling convective motions in the asthenosphere can break through the crust, creating mid-oceanic rifts. On either side of a mid-oceanic rift, the motions of the underlying convective cells drag the ocean plates apart (Hess 1962). This phenomenon is referred to as seafloor spreading. The low pressures within the vicinity of the rift allow the upwelling material to liquefy and become magma, leading to the creation of undersea

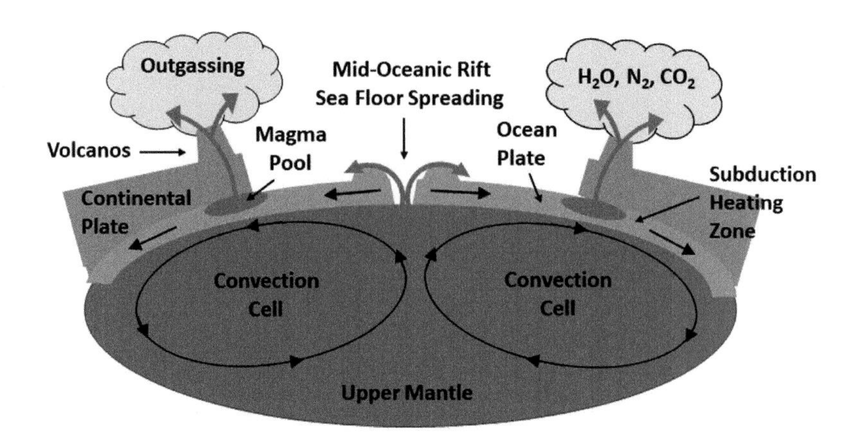

FIGURE 5.25 Plate tectonic. The opposing motion of convection cells in the upper mantle (i.e., asthenosphere) rips the Earth's crust apart where it is thinnest creating what will become a mid-oceanic rift. Upwelling magma from the mantle pours through the rift. As the basaltic ocean plates spread they encounter lower-density continental plates. As they slide by one another heat generated from friction creates pools of high-pressure magma that crack the crust. The cracks provide channels by which magma can find its way to the surface and create volcanoes.

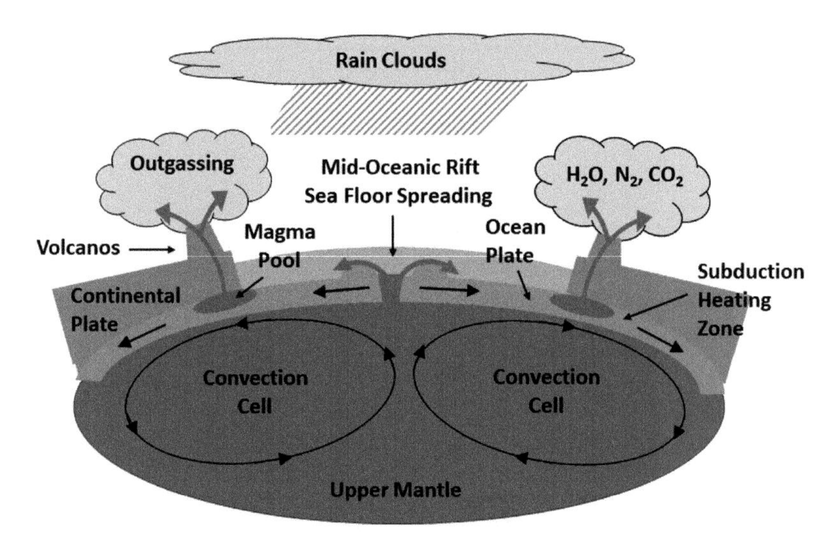

FIGURE 5.26 Origin of the Earth's atmosphere. Still hot from creation, the surface of the primordial Earth was a violent place, covered with countless volcanoes each spewing out prodigious amount of CO_2, N_2, and H_2O. After millions of years volcanic outgassing had increased the atmospheric pressure to the point where rain clouds could form and the ocean's basins began to fill with water.

volcanoes along its length. The spreading motion of the seafloor will ultimately cause it to crash into a continental plate. When this occurs the denser oceanic plate dives under the lighter continental plate. The region where this occurs is called the subduction zone (see Figures 5.25 and 5.26). Water may play an important role as a lubricant that facilitates the passage of one plate over another (Tikoo and Elkins-Tanton 2017).

Frictional heating associated with the passing of one plate over another creates subterranean pools of magma. If the pressure in the pools is sufficiently high, cracks will form in the brittle continental plates creating channels through which high pressure magma can find its way to the surface, leading to the creation of volcanoes. The "Ring of Fire" associated with the collision of the Pacific plate with coastal land masses is a recent example of this phenomenon.

The study of the geologic processes and effects associated with the movements of oceanic and continental plates is called plate tectonics. It is a unifying theory in geology that explains many observed phenomena, including continental drift, volcanoes, and earthquakes. The idea that the continents are like puzzle pieces that have drifted apart due to continental drift was first suggested by the Dutch cartographer Abraham Ortelius upon examining the appearance of the world map in his 1596 edition of *Thesaurus geographicu*. It was later independently developed by Otto Ampferer in 1906 and more famously by Alfred Wegener in 1912. As we have seen is often the case, an idea which we now know is correct was largely rejected at the time. It was not until the discovery of matching paleomagnetic strips on either side of the Atlantic mid-oceanic rifts that the theory of seafloor spreading and plate tectonics began to be widely accepted. Paleomagnetic stripes form when iron crystals within magma emerging from a mid-oceanic rift align themselves with the prevailing magnetic field (like little compass needs) before the magma cools and freezes them in place. Not too surprisingly, the first paper submitted describing the use of paleomagnetic stripes as evidence for seafloor spreading was initially rejected by the journal *Nature* (Vine and Matthews 1963). The alternating orientation of the iron crystals in the paleomagnetic stripes is itself interesting, in that it indicates that the polarity of the Earth's magnetic field swaps polarity every 2,000–12,000 years (Glatzmaier and Cole 2015). Short-term reversals due to field reversals due to changing currents in the liquid outer core (but not in the solid inner core) can happen on ~500-year timescales (Gubbins 1999). During field reversals the strength of the Earth's magnetic field drops to ~10% of its full value, effectively "lowering the shields" to the solar wind. Such reversals have happened many times over the Earth's geological history, apparently not adversely affecting life. But with our civilization's dependence on satellite communications and a stable power grid, the next occurrence will likely not go unnoticed.

One reality of plate tectonics is that the current appearance of the Earth is transitory. The diameter of the Earth is just under 8,000 miles or 12,742 km. The circumference of the Earth is then ~40 million cm. If the tectonic plates move at about 1 cm per year, this means the surface of the Earth recreates itself every 40 million years or so. There is ample evidence for this game of tectonic billiards in the fossil (Benton 2005) and geological record (Murck and Skinner 1999). Occasionally, the land masses happen to all come together and form a supercontinent. The last such supercontinent was Pangea. It existed during the Paleozoic and Mesozoic eras and started to break up during the late Triassic and early Jurassic eras (Rogers and Santosh 2004). This is why fossils of some dinosaur species have been found spread across continental land masses that are today separated by oceans.

5.5.6 Origin of Earth's Atmosphere

Another powerful manifestation of plate tectonics is the existence and composition of the Earth's atmosphere. After its formation from the collision of planetesimals and the subsequent giant impact resulting in the formation of the Moon, the Earth had little or no atmosphere. Whatever atmosphere the Earth may have accumulated beforehand was stripped off by the impact. The rock vapor left behind after the impact is expected to have condensed out on to the surface within a few thousand years (Cuk and Stewart 2012). The principal source for the Earth's primordial atmosphere was outgassing from volcanoes (see Figures 5.25, 5.26, and 5.27). The temperature of magma, up to 2,440 °F (1,292 °C) is sufficient to release volatiles trapped within rock. The volatiles released were primarily carbon dioxide (CO_2), molecular nitrogen (N_2), water vapor (H_2O), with smaller amounts of hydrochloric acid (HCL), methane (CH_4), ammonia (NH_3), and sulfur compounds. During this early epoch the interior of the Earth was still quite hot and the crust relatively thin, which resulted in its surface being covered with countless volcanoes. After ~0.5 billion years the collective outgassing of these volcanoes had built-up the atmospheric pressure to a level sufficient to allow the formation of clouds. An atmosphere with this composition, i.e., containing no free oxygen and mildly acidic, is classified as reducing. Rainfall from these clouds filled the low-lying, basaltic regions of the Earth's crust with water to create oceans, lakes, and rivers. Evidence for these conditions prevailing for the first ~ 2 billion years of Earth's existence comes to us from the geologic record where layers of sulfide minerals (evidence for a reducing environment) and chert or fine-grained quartz are not found in rocks formed in the last ~2 billion years ago. Late cometary impacts are expected to have contributed to the Earth's water supply. However, recent evidence from the measurement of the hydrogen content and isotopic ratios within enstatite chondrite (EC) meteorites suggests the planetesimals that formed the proto-Earth had sufficient water content to supply the Earth's oceans several times over (Piani et al. 2020).

5.5.7 Primordial CO_2 Cycle

Initially, the atmospheric CO_2 level was between 10 and 200 times higher than what it is today. CO_2 is a greenhouse gas, i.e., it passes ultraviolet light but blocks the passage of the thermal radiation produced when the light strikes surfaces (like the glass in a greenhouse or the windshield of a car). Such high CO_2 levels would drive up the planet's surface temperature, promoting evaporation and the creation of powerful storms, similar to what we are seeing today with global warming, but much more severe. The torrential rainfall promoted the erosion of the crust. Through the process of carbon mineralization, CO_2 combined with exposed igneous and metamorphic rock, reducing its atmospheric concentration. Some of this CO_2-enriched material was washed into the oceans where it accumulated on the seafloor. Another pathway to leech CO_2 out of the air occurred at the boundary between the oceans and the atmosphere. There, H_2O combined with CO_2 to create carbonic acid (H_2CO_3), a liquid at ambient temperature (25 °C) and pressure (1 atm). When the carbonic acid encountered calcium that had been dissolved in the ocean water from surface erosion, $CaCO_3$, i.e., calcium carbonate, is formed. Being a solid the $CaCO_3$ sinks and slowly accumulates on the seafloor. Some fraction of the trapped CO_2 on the ocean floor is ultimately released back into the atmosphere through volcanic outgassing as a result of plate tectonic

FIGURE 5.27 Volcanic outgassing. The eruption of Mt. St. Helens on May 18, 1980, illustrates the ability of volcanoes to pump prodigious amounts of gas and debris into the Earth's atmosphere. Mt. St Helens is the result of the Pacific plate colliding with the North American plate and is a part of the Pacific Ring of Fire. Credit: USGS, https://pubs.usgs.gov/gip/msh/climactic.html

processes (see Figure 5.27). The interrelationship between CO_2 level, rainfall, and plate tectonics creates a negative feedback loop that stabilizes the Earth's temperature over time. If the CO_2 level gets too high, it rains more, which washes CO_2 out of the atmosphere until the temperature drops and the rainfall slows. The CO_2 level will then slowly increase through volcanic outgassing until it once again triggers rainfall until the excess CO_2 is washed out. This is the primordial CO_2 cycle and was essential to creating an environment conducive to the origin of life on Earth. Not all planets are so lucky.

The range of distances from a star over which a planet could potentially have liquid water on its surface is called the habitable zone. In our solar system, Venus is on the inner edge of the habitable zone and Mars is on the outer. Both worlds show strong evidence for the presence of surface water soon after they formed, but within a billion years it was gone. In the case of Venus, it is believed the CO_2 level increased to the point that there was a runaway green house, resulting in the evaporation of all surface water (Kasting 1988). Since water is also a greenhouse gas, this exacerbated the problem even further. Today the surface temperature of Venus is a staggering ~900 °F (475 °C) and the atmospheric pressure is a crushing ~90 atm. In the case of Mars, ~99% of its atmosphere appears to have been stripped away by the solar wind as a result of not having a protective magnetic field (Jakosky et al. 2017). Neither Venus nor Mars has plate tectonics, therefore, were unable to establish a CO_2 cycle. Even if they had, being at the edge of the habitable zone may have led to climatic extremes beyond what the CO_2 cycle could handle. The cautionary tale of Venus and Mars suggests that there is a higher probability of finding conditions suitable for life on planets located near the middle of a star's habitable zone. We will discuss habitable zones further in Chapter 8.

SUMMARY

In this chapter, we discussed the insights, discoveries, and observations that have led to our current understanding of how the solar system evolved from a rarefied interstellar gas cloud ~4.5 billion years ago. We also learned how an Earth-like planet forms and what geological processes determine physical conditions in its interior and on its surface. The conditions on the surface of the primordial Earth may seem inhospitable, but, as we shall see in the next chapter, they made the Earth ripe for the origin of life.

REVIEW QUESTIONS

1) Who first provided a qualitative description of planet formation?
2) Who first provided a quantitative theory of planet formation? What was it based on?

   What was wrong with it?
3) How was the planet formation theory of 2) ultimately fixed?
4) Who was the first to characterize a protoplanetary disk from observations? What kind of data was used?
5) Who first showed protoplanetary disks obey Kepler's Laws? How was it done and what kind of telescope was used?
6) Describe the type of morphologies observed in protoplanetary disks.
7) What evidence is there for planet formation in protoplanetary disks? Are there any analogs in the solar system?
8) What is the planetesimal theory of planet formation?
9) What evidence do is there to back up the planetesimal theory?
10) What are examples of left-over planetesimals?
11) Where are most left-over planetesimals located? How did they get there?
12) Did our solar system always have the same planets it does today?
13) How long did it take for the planets to form in the protoplanetary disk?
14) Why do Jovian planets form in the outer solar system?
15) How can we date the surfaces of planets just by looking at them?
16) What and when was the Late Heavy Bombardment?
17) Draw a picture of the Earth's cross section and identify key features.
18) What geological process is responsible for heavier things settling at the Earth's center?
19) How can seismic waves be used to probe the internal structure of the Earth?
20) When was the first seismometer demonstrated?
21) If you were able to see the core of the Earth, what color would it be? Why?

22) What is the leading theory for the formation of the Moon?
23) What is the origin of the Earth's magnetic field?
24) What is the origin of the northern and southern lights?
25) What does the Earth's magnetic field protect us from?
26) What is plate tectonics?
27) How was Earth's primordial atmosphere produced?
28) What are paleomagnetic stripes?
29) What is the CO_2 cycle? Why is it a good thing?
30) Why are the atmospheres of Venus and Mars so different from ours?
31) Is there any evidence that water has existed on the surfaces of planets other than the Earth?

REFERENCES

Alexander, Amir. 1998. "Lunar Maps and Coastal Outlines: Thomas Hariot's Mapping of the Moon." *Studies in History & Philosophy of Science* 29 (3): 345–68.

Andrews, S. 2021. "The Structures of Protoplanetary Disks." *Physics Today* 74 (8): 37.

Andrews, S. and Other. 2018. "The Disk Substructures at High Angular Resolution Project (DSHARP): I. Motivation, Sample, Calibration, and Overview." *The Astrophysical Journal Letters* 869: L41.

Baker, G. 1983. "An 18th Century Cosmologist." *Physics Teacher*, October Issue: 44.

Balbus, S. A., and J. F. Hawley. 1991. "A Powerful Local Shear Instability in Weakly Magnetized Discs 1. Linear Analysis", *The Astrophysical Journal* 376: 214.

Ball, W. W. R. 1908. *A Short Account of the History of Mathematics*. London and New York: Macmillan.

Beckers, J. M. 1993. "Adaptive Optics for Astronomy: Principles, Performance, and Applications." *Annual Review of Astronomy & Astrophysics* 31 (1): 13–62.

Beckwith, S., A., Sargent, N., Scoville, C., Masson, B., Zuckerman, and T., Phillips. 1986. "Small Scale Structure of the Circumstellar Gas of HL Tauri and R. Monocerotis." *Ap.J.*, 309: 755.

Beckwith, S. V. W., and A. I. Sargent. 1993. "The Occurrence and Properties of Disks around Young Stars" In *Protostars and Planets III*, edited by E. H. Levy, and J. I. Lunine, 521. Tucson: University of Arizona Press.

Benton, M. J. 2005. *Vertebrate Palaeontology*, 3rd ed., 25. Wiley-Blackwell: Hoboken, NJ.

Blandford, R., and D. G. Payne. 1982. "Hydrodynamic Flows from Accretion Discs and the Production of Radio Jets." *Monthly Notices of the Royal Astronomical Society* 199 (4): 883.

Boss, A. 2017. "'Triggering Collapse of the Presolar Dense Cloud Core and Injecting Short-lived Radioisotopes with a Shock Wave. V. Nonisothermal Collapse Regime'." *The Astrophysical Journal* 844: 113.

Cameron, A. G. W. 1978. "Physics of the Primitive Solar Accretion Disk". *Moon & the Planets* 18 (1): 5.

Chamberlin, T. C. 1901. "On a Possible Function of Disruptive Approach in the Formation of Meteorites, Comets, and Nebula", *The Astrophysical Journal* 14: 17.

Clerke, Agnes Mary. 1911. "Laplace, Pierre Simon." In *Encyclopædia Britannica*, edited by Hugh Chisholm, 11th ed., Vol. 16, 200–2. Cambridge University Press: Cambridge, UK.

Cohen, M. 1983. "HL Tau and its circumstellar disk", *The Astrophysical Journal Letters* 270: L69.

Cuk, M., and S. T. Stewart. 2012. "Making the Moon from a Fast-Spinning Earth: A Giant Impact Followed by Resonant Despinning." *Science* 338 (6110): 1047–52.

Dormand, J. R., and M. M. Woolfson. 1989. *The Origin of the Solar System; the Capture Theory.* Chichester: Ellis Horwood.

Dyck, H. M., and R. R. Howell. 1982. "Speckle Interferometry of molecular cloud sources at 4.8 microns", *The Astrophysical Journal* 87: 1223.

Galileo, G. 1610. *Sidereus Nuncius*, translated by A. Van Helden 1989. Chicago: The University of Chicago Press.

Gillispie, C., and I. Grattan-Guinness. 2000. *Pierre-Simon Laplace, 1749–1827.* Princeton: Princeton University Press.

Glatzmaier, G. A., and R. S. Coe. 2015. "Magnetic Polarity Reversals in the Core." *Treatise on Geophysics*, 2nd Edition 8, 279–95.

Gomes, R., H. F. Levison, K. Tsiganis, and A. Morbidelli. 2005. "Origin of the Cataclysmic Late Heavy Bombardment Period of the Terrestrial Planets." *Nature* 435 (7041): 466–9.

Gubbins, David. 1999. "The Distinction between Geomagnetic Excursions and Reversals." *Geophysical Journal International* 137 (1): F1–F4.

Herbig, G. H. 1968. "The Structure and Spectrum of R Monocerotis", *The Astrophysical Journal* 152: 439.

Herbig, G. H., and K. N. Rao. 1972. "Second Catalog of Emission-Line Stars of the Orion Population", *The Astrophysical Journal* 174: 401.

Hess, H. H. 1962. "History of Ocean Basins." In *Petrologic Studies: A Volume in Honor of A. F. Buddington*, edited by A. E. J. Engel, Harold L. James, and B. F. Leonard, 599–620. Boulder, CO: Geological Society of America, November 1.

Jacobson, Seth A., A. Morbidelli, S. N. Raymond, D. P. O'Brien, K. J. Walsh, and D. C. Rubie. 2014. "Highly Siderophile Elements in Earth's Mantle as a Clock or the Moon-Forming Impact." *Nature* 508 (7494 (April): 84–7.

Jakosky, B. M., M. Slipski, M. Benna, P. Mahaffy, M. Elrod, R. Yelle, S. Stone, and N. Alsaeed. 2017. "Mars' Atmospheric History Derived from Upper-Atmosphere Measurements of ^{38}Ar/^{36}Ar." *Science* 355 (6332): 1408.

Jeans, J. H. 1917. "The Equations of Radiative Transfer of Energy", *Memoirs of the Royal Astronomical Society* 77: 186.

Jones, J. H. 1998. "Tests of the Giant Impact Hypothesis." *Lunar and Planetary Science.* Origin of the Earth and Moon Conference. Monterey, California.

Johnston, I. 2008. "Universal Natural History and Theory of the Heavens." *Richer Resources Publications.* Arlington, Va.

Kant, Emanuel. 1755. "Universal Natural History and Theory of the Heavens or an Essay on the Constitution and the Mechanical Origin of the Entire Structure of the Universe Based on Newtonian Principles." Translated by Ian Johnston. Vancouver Island University Mécanique Celeste.

Kasting, J. F. 1988. "Runaway and Moist Greenhouse Atmospheres and the Evolution of Earth and Venus." *Icarus* 74 (3): 472–94.

Klahr, H., and P. Bodenheimer. 2003. "'Turbulence in Accretion Disks: Vorticity Generation and Angular Momentum Transport via the Global Baroclinic Instability'." In *Planetary Systems in the Universe- Observations, Formation, and Evolution.* Proceedings IAU Symposium No. 202, p. 350.

Kronk, Gary W. 1999. "Cometography." Vol. 1: Ancient-1799, Cambridge University Press, p. 14. and perhaps earlier. Rincon, Paul (10 September 2010). "Halley's Comet 'was Spotted by the Ancient GREEKS'." *BBC News.*

Lüst, R. 1952. "Die Entwicklung einer um einen Zentralkörper rotierenden Gasmasse. I. Lösungen der hydrodynamischen Gleichungen mit turbulenter Reibung ", Zeitschrift für Naturforschung.". A 7(1), 87-98.

Laplace, P. S. de. 1796. *Exposition du Systeme du Monde*. Paris: Imprimerie Cercle-Social.

Laplace, Pierre Simon, marquis de, 1749–1827; Bowditch, Nathaniel, 1773–1838; Bowditch, N. I. (Nathaniel Ingersoll), 1805–1861.

Lauretta, D. S., D. N. DellaGiustina et al. 2019. "The Unexpected Surface of Asteroid (101955) Bennu." *Nature*. https://doi.org/10.1038/s41586-019-1033-6.

Louvet, F., C. Dougados, S. Cabrit, D. Mardones, F. Menard, B. Tabone, C. Pinte, and W. R. F. Dent. 2018. "The HH30 edge-on T Tauri star A rotating and precessing monopolar outflow scrutinized by ALMA." *A&A* 618: A120.

Lyttleton, R. A., 1960. "Dynamical Calculations Relating to the Origin of the Solar System," *MNRAS* 121(6): 551.

Massey, R. 2009. https://web.archive.org/web/20130627230355/http://www. ras.org.uk/news-and-press/68-news2009/1641-celebrating-thomas-harriot-the-worlds-first-telescopic-astronomer-ras-pn-0947.

Moulton, F. R. 1905. "On the Evolution of the Solar System", *The Astrophysical Journal* 22: 165.

Murck, Barbara W., and Brian J. Skinner. 1999. *Geology Today: Understanding Our Planet, Study Guide*. Wiley: Hoboken, NJ. ISBN 978-0-471-32323-5.

Murray, C. D., and S. F. Dermott. 1999. *Solar System Dynamics*, 184. Cambridge University Press.

Neugebauer, G., H., Habing, H., van Duinen et al. 1984. "The Infrared Astronomical Satellite (IRAS) Mission", *Ap.J.*, 278, L1.

Newton, Isaac. 1687. *Philosophiae Naturalis Principia Mathematica [The Mathematical Principles of Natural Philosophy]*.

O'Connor, John J., and Edmund F. Robertson. *Pierre-Simon Laplace, MacTutor History of Mathematics archive. University of St Andrews*. Accessed August 25, 2007.

Paniello, Randal C., James M. D. Day, and Frédéric Moynier. 2012. "Zinc Isotopic Evidence for the Origin of the Moon." *Nature* 490, no. 7420 (October): 376–9.

Piani, P., Y. Marrocchi, T. Rigaudier, L. Vacher, D. Thomassin, and B. Marty. 2020. "Earth's Water May Have Been Inherited from Material Similar to Enstatite Chondrite Meteorites." *Science* 369 (6507): 1110.

Powell, C. "What Did the Moon Look Like from Earth 4 Billion Years Ago?" July. https://www.forbes.com/sites/quora/2018/07/11/what-did-the-moon-look-like-from-earth-4-billion-years-ago/?sh=685198bb1151.

Rogers, J. J. W., and M. Santosh. 2004. *Continents and Supercontinents*. Oxford: Oxford University Press, 146.

Rosetta. 2015. "Astronomy & Astrophysics Special Issue 'Rosetta Mission Results Pre-perihelion'." *A&A* 583.

Saal, A. E., E. H. Hauri, J. A. Van Orman, and M. J. Rutherford. 2013. "Hydrogen Isotopes in Lunar Volcanic Glasses and Melt Inclusions Reveal a Carbonaceous Chondrite Heritage." *Science* 340 (6138): 1317–20.

Sargent, A. I., and S. V. W. Beckwith. 1991. "'The Molecular Structure Around HL Tauri'." *The Astrophysical Journal* 382: L31–L5.

Shakura, N. I., and R. A. Sunyaev. 1973. "Black holes in binary systems: Observational Appearance", *A&A*. 24: 337.

Spitzer, L. 1939. "The Dissipation of Planetary Filaments", *The Astrophysical Journal* 90: 675.

Stevenson, D. J. 1987. "Origin of the Moon–The Collision Hypothesis." *Annual Review of Earth & Planetary Sciences* 15 (1): 271–315.

Strom, S. E., S. Edwards, and M. F. Skrutskie. 1993. "Evolutionary Time Scales for Circumstellar Disks Associated with Intermediate-Type and Solar-Type Stars", In *Protostars and Planets III*, edited by E. H. Levy, and J. I. Lunine, 837. Tucson: University of Arizona Press.

Swedenborg, Emanuel. 1734. *(Principia) Latin: Opera Philosophica et Mineralia (English: Philosophical and Mineralogical Works). I.*

Swedenborg, Emanuel. 1907. *Heavenly Arcana* (or *Arcana Coelestia*), 1749–58 (AC). 20 vols. Rotch Edition. New York: Houghton, Mifflin and Company, in *The Divine Revelation of the New Jerusalem* (2012), n. 1799(4).

Tikoo, Sonia M., and Linda T. Elkins-Tanton. 2017. "The Fate of Water within Earth and Super-Earths and Implications for Plate Tectonics." *Philosophical Transactions. Series A* 375 (2094): 20150394. https://doi.org/10.1098/rsta.2015.0394.

Vine, F. J., and D. H. Matthews. 1963. "Magnetic Anomalies Over Oceanic Ridges." *Nature* 199 (4897): 947–9.

Walker, C. K., C. J., Lada, E. T., Young, M., Margulis, and B., Wiking. 1985. "An Unusual High Velocity Molecular Outflw in the Rho Ophiuchu Cloud," *BAAS* vol. 17, 835.

Walker, C. K., C. J., Lada, E. T., Young, Maloney P. R., Wilking B. A. 1986. "Spectroscopic Evidence for Infall around an Extraordinary Source in Ophiuchus," *Ap.J. Letters* 309: L47.

Walker, C. K., J. E. Carlstrom, and J. H. Bieging. 1993. "'The IRAS 16293 - 2422 Cloud Core: A Study of a Young Binary System'." *The Astrophysical Journal* 402: 655.

Whittaker, Edmund. 1949. "Laplace." *Mathematical Gazette* 33 (303): 1–12.

Williams, James G., and Dale H. Boggs. 2016. "Secular Tidal Changes in Lunar Orbit and Earth Rotation." *Celestial Mechanics & Dynamical Astronomy* 126 (1): 89–129.

Woodward, R. S. 1891. "Review of Tisserand's *Mecânique Céleste.*" *Annals of Mathematics* 6 (2): 49–56.

Woolfson, M. M. 1964. "The Capture Theory of the Origin of the Solar System", *Proceedings of the Royal Society of London Series A* 282 (1391): 485.

Woolfson, M. 1993. "The Solar System - its Origin and Evolution", *Q. J. R. astr. Soc.*, 34: 1–20.

CHAPTER 6

The Origin of Life

PROLOGUE

Of all the factors in the Drake Equation perhaps the least understood is f_L, the fraction of Earth-like planets on which life originates. Looking around, we see that life permeates our planet. Indeed, there are over 10 million known species of life on Earth. It can be found on the highest mountains, at the bottom of the deepest oceans, and in the hottest and coldest places. The fossil record tells us life appeared on Earth in its simplest form within a billion years of formation. As we shall see, the fabric of life shares a common thread that can be traced back to these early times. The environs of the early Earth are in themselves not unique and are expected to be found on trillions of planets and moons throughout the cosmos. Although we do not yet understand exactly how it happened, these facts suggest the origin of life may not be unique to the Earth but simply a byproduct of the physical evolution of the Universe.

6.1 INTRODUCTION

What were the conditions like on early Earth? As we learned in Chapter 5, the world had recently been created from the collisions of planetesimals leaving behind a thin, newly formed rocky crust enveloping a magma sea far warmer and active than the interior of today's Earth. As is the case on most terrestrial worlds, numerous volcanoes pock-marked the surface, each belching out copious amounts of carbon dioxide (CO_2), nitrogen (N_2), and water (H_2O) (see Figure 6.1). The Earth's gravity was sufficiently strong to hang on to these relatively heavy molecules, and they began to accumulate to form the Earth's primordial atmosphere. The Earth's distance from the Sun was far enough that gaseous water could condense to form liquid water droplets and rain clouds but not so far from the Sun that water vapor from volcanoes would freeze. And rain it did, for millions of years. Thunderstorms were likely common, sending bolts of lightning across the sky. Rain water slowly filled the low-lying regions of the crust while washing out a significant fraction of the CO_2 from the atmosphere. Lower CO_2 levels kept the Earth from experiencing a runaway greenhouse effect like is now occurring on our sister planet Venus. Enough CO_2 was left behind that it could serve as a winter blanket to keep most of the Earth above the freezing

DOI: 10.1201/9781315210643-6

FIGURE 6.1 Early earth. This illustration depicts the violent conditions on Earth within a few 100 million years after its formation, about 4.44 billion years ago. The thin newly formed crust hosts countless volcanoes outgassing the nitrogen, carbon dioxide, and water vapor that constituted our early atmosphere. The water vapor cooled to form clouds and thunderstorms that rained down on the Earth, yielding the watery world we have today. This geologically produced water was augmented by the water from cometary bodies that continued to pelt the planet in large numbers. Image: NASA.

point of water. During these early days, there were still numerous planetesimals within the inner solar system. Often passing meteors and comets would be captured by Earth's gravitational field and sent plummeting to the surface adding to the carnage. The water, rock, and minerals from these bodies became part of Earth's treasure trove. If we were to visit the Earth during this time period, it would be quite inhospitable, and we would need breathing apparatus. But within the newly formed oceans and on the rocky coastlines surrounding them, conditions were ripe for the origin of life. Initially, only organic molecules would be found there. Liquid water acted as the medium through which these molecules could interact and become increasingly complex. This is the era of chemical and molecular evolution on the Earth. Based upon fossil evidence in ancient rocks, simple, unicellular forms of life appear to have originated on Earth within ~600 million years of its formation, not long after the crust cooled and solidified. For the first 3 billion years of its 4.54 billion years of existence, only single-celled lifeforms inhabited our world. In this chapter, we will explore how these first lifeforms may have originated in the hostile environment of the primordial Earth.

6.2 THE BUILDING BLOCKS OF LIFE

All life on Earth makes use of four types of molecules: proteins, nucleic acids, lipids, and carbohydrates (see Figure 6.2). In general, proteins act as structural elements or "bricks" and, when acting as enzymes, can help regulate cellular processes. Nucleic acids act as the "libraries" of knowledge for cells. Lipids are long chains of fatty compounds that do not dissolve in water and, like proteins, serve important roles in the structure and function of cells. Carbohydrates are mostly sugars and serve as an energy source for cells.

6.2.1 Proteins

Proteins are polymers. A polymer is a macromolecular structure consisting partly or entirely of a large number of similar units bonded together. The units are often referred to as monomers. The proteins associated with life on Earth contain 20 different types of monomers, also known as amino acids. These amino acids can be combined in a multitude (e.g., 2^{20}) of different ways to make a large variety of proteins. The simplest of all amino acids is glycine (Figure 6.3 (*Left*)). What makes amino acids special is their ability to link together by forming peptide bonds. A linear chain of amino acids is called a polypeptide. A protein contains at least one such long polypeptide. Each amino acid begins with an amine group of atoms composed of one nitrogen and two hydrogen atoms that can be thought of as a "hook". The amino acid ends with a carboxyl group, composed of one carbon, two oxygen, and one hydrogen atom that can be thought of as an "eyelet". In between the two groups is a chain of one or more carbon atoms with bonds to neighboring atoms (e.g., sulfur, hydrogen, oxygen) that gives the amino acid its functional personality. If one makes the analogy of a polymer being a train composed of cars, it is the amine and carboxyl groups that tie the cars together. The cargo in a car is the chain of atoms and associated

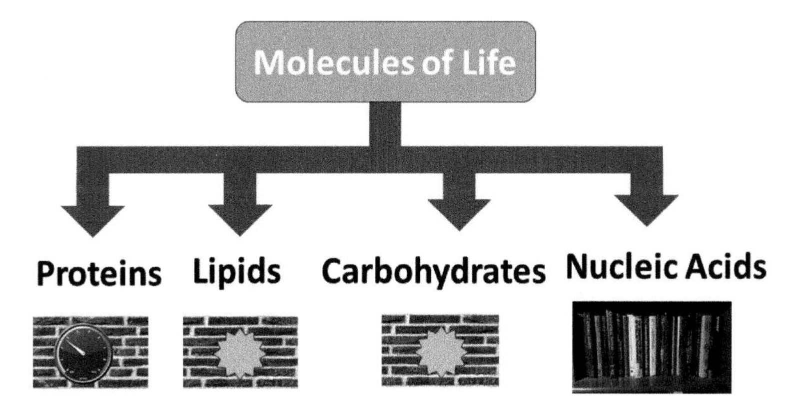

FIGURE 6.2 Molecules of life. Within cells proteins serve as building blocks and regulate activity. Depending on the organism, lipids and carbohydrates can be either building blocks and/or a source of energy. Nucleic acids are where the information needed to create and maintain life is stored.

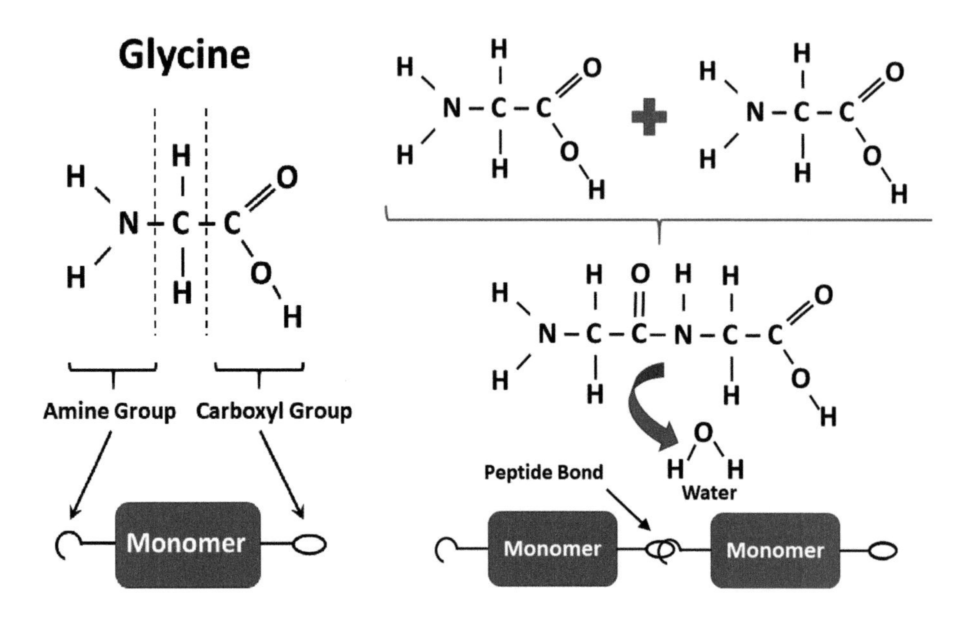

FIGURE 6.3 *Left*: Chemical structure of glycine. Amino acids, like glycine, are referred to as monomers. They have an amine group on one end and a carboxyl group on the other. These groups are key to forming the peptide bonds that link monomers together to form polymers. *Right*: Formation of peptide bonds. Proteins are polymers. Polymers are formed by linking together monomers. The chemical link between monomers is a peptide bond. When a peptide bond is formed an excess water molecule is released. Likewise, water molecules can be used to disrupt (or dissolve) peptide bonds. After Evans 1996.

atoms between the two groups. The formation of a peptide bond is an endothermic reaction, i.e., energy is required to make it happen. In the case of train cars, they are rammed together. In the case of amino acids, they can be linked together through collisions (like train cars) or by applying heat or attaching an energy-rich molecule, such as adenosine triphosphate (ATP).

Referring to Figure 6.3 (*Right*), it can be seen that when a peptide bond is formed, the participating amine group loses a hydrogen atom and the carboxyl group loses an oxygen and hydrogen atom. These three byproduct atoms together form water, H_2O. Likewise, if you want to break a peptide bond, the addition of water molecules can work to dissolve the peptide bond between two or more amino acids. Anytime you have soaked dishes in a sink to loosen dried-on food, you have witnessed this phenomenon. Indeed, this process of making and breaking peptide bonds with water is occurring millions of times a day within the cells of your body. It underpins biochemistry and is fundamental to all life on Earth. In most instances, without water existing in the narrow 100°C temperature range between its boiling and freezing points, biochemistry to a large extent ceases; at lower temperatures, water freezes and metabolism stops, and at higher temperatures polymers are torn asunder. These are the reasons why frozen food does not spoil and boiling things in water is an effective sterilization technique.

Most proteins are not 2D as Figure 6.3 suggests but are actually complex 3D structures whose mechanical properties play an important role in how it is used, for instance, in the creation of a permeable cell wall. Proteins are also essential to metabolism, serving as enzymes that help speed up (i.e., catalyze) biochemical reactions in cells. Once formed, a protein can exist for minutes or years within an organism before it is expelled or recycled for its amino acid building blocks.

6.2.2 Nucleic Acids

Essential to the existence of life are the nucleic acids within whose structure the detailed instruction manual for the fabrication and operation of a cell is contained. Indeed, a virus, arguable the simplest form of life on Earth, is little more than a nucleic acid wrapped in a protein coating (see Figure 6.4). In modern organisms, nucleic acids come in two forms, ribonucleic acid (RNA) and deoxyribonucleic acid (DNA). A photograph of a DNA molecule is shown in Figure 6.5. DNA appears as a twisted ladder, while RNA resembles half of a ladder that has been sawed down the middle. The rungs of the ladder are made of base molecules. There are five types of bases that are utilized by DNA and RNA by life on Earth: adenine (A), guanine (G), cytosine (C), thymine (T), and uracil (U). More specifically, DNA uses A, G, C, and T, while RNA uses A, G, C, and U. It is in the arrangement, or sequencing, of this handful of bases through which information is conveyed. The sides of the ladder are long chains of alternating phosphate (P) and sugar (S) molecules. In the case of DNA deoxyribose sugars are used. In RNA ribose sugars are used. The difference being deoxyribose sugar has one less oxygen atom. Unlike RNA, the DNA molecule has (see Figure 6.5; also Figure 6.10) two sides (or guard rails) of phosphates and sugars that help protect the base sequences from damage and thereby safeguard the knowledge they contain.

The molecular structure of sugars, phosphates, and bases are in themselves complex (see Figures 6.6 and 6.7). So, to facilitate the discussion, we will adopt the symbolic

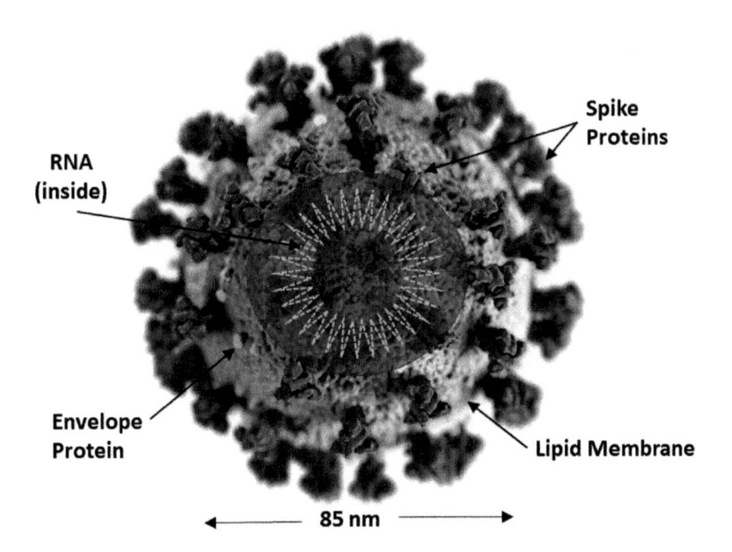

FIGURE 6.4 Coronavirus. Consisting of little more than a strand of nucleic acid and a protective protein shell, viruses are arguably the simplest form of life on Earth. A virus contains none of the cellular machinery required to reproduce themselves. They replicate (perpetuate) themselves by commandeering the machinery of a host cell. The RNA within COVID-19 contains almost 30,000 nucleotides. Adapted from CDC: Public Domain.

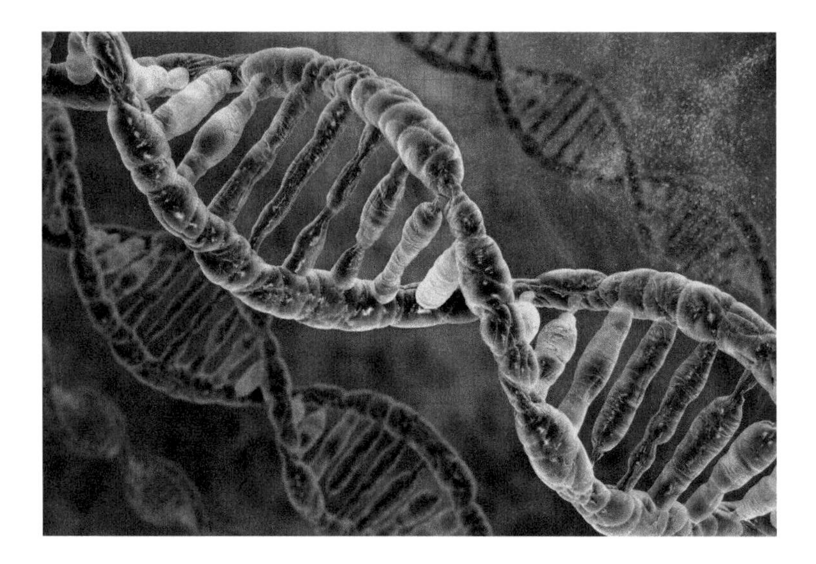

FIGURE 6.5 The DNA molecule. The large DNA molecule is a composite of several smaller molecules. The "sides" of the latter are composed of alternating phosphate and sugar molecules. The "rungs" of the latter are composed of pairs of base molecules. Image Credit: Shutterstock.

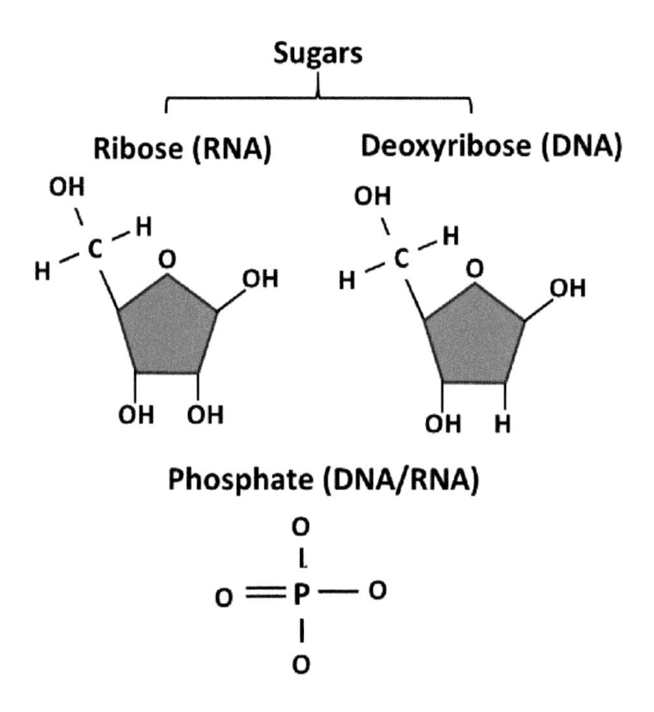

FIGURE 6.6 Chemical structures of sugars and phosphates: The "sides" of DNA and RNA are made of alternating molecules of phosphates and sugars. The "D" in DNA comes from the fact it uses deoxyribose sugars, while the "R" in RNA comes from its use of ribose sugars. They appear the same except the deoxyribose sugar has one less oxygen atom. The blue region indicates where electrons are being shared between the composite atoms.

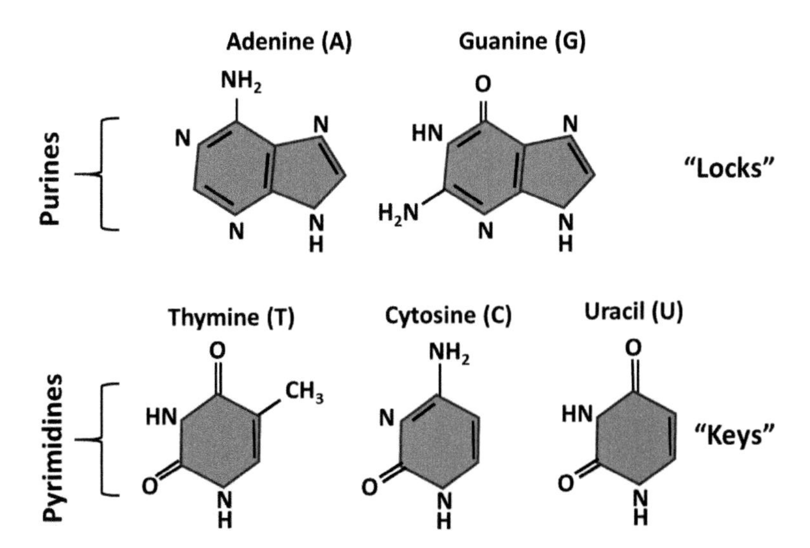

FIGURE 6.7 Chemical structures of bases: the "rungs" of DNA and RNA are made of base pairs. There are five types of bases, two of which are classified as purines (A, G) and three classified as pyrimidines (T, C, U). A pair consists of one of each type, with a purine serving as the "lock" and the pyrimidine as the mating "key". Allowable base pairs are then AT, AC, AU, GT, GC, and GU. In RNA thymine (T) is used, while uracil (U) is used in RNA.

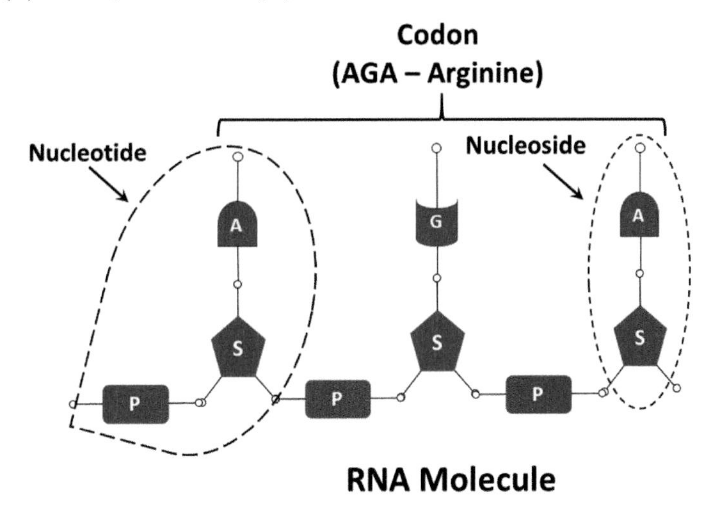

FIGURE 6.8 RNA structure: The side of the RNA ladder is composed of alternating phosphates and sugars. The rung of the ladder consists of bases. Three bases together form a "codon" of information. The genetic code of Table 6.1 is composed of 64 such codons. The letter sequence of a codon is sufficient to designate one of 20 amino acids used by lifeforms on Earth. The code is common to all life and suggests all life has a common origin. In genetics the combination of a phosphate, sugar, and base is referred to as a nucleotide; the combination of a base and sugar as a nucleotide.

nomenclature of Figure 6.8. In genetics a sugar and base together are referred to as a nucleoside. If a phosphate is added to this structure it is referred to as a nucleotide. It only takes three bases (or nucleotides) to specify which of the 20 amino acids at a given time is to be made by a cell. Table 6.1 lists the standard genetic codes for RNA. Since a codon consists of three bases, the number of entries is 2^3, or 64. Notice that there is redundancy in the code,

TABLE 6.1 The Genetic Code

	Second Letter			
Base	**U**	**C**	**A**	**G**
U	UUU: Phenylalanine	UCU: Serine	UAU: Tyrosine	UGU: Cysteine
	UUC: Phenylalanine	UCC: Serine	UAC: Tyrosine	UGC: Cysteine
	UUA: Leucine	UCA: Serine	UAA: Stop	UGA: Stop
	UUG: Leucine	UCG: Serine	UAG: Stop	UGG: Tryptophan
C	CUU: Leucine	CCU: Proline	CAU: Histidine	CGU: Arginine
	CUC: Leucine	CCC: Proline	CAC: Histidine	CGC: Arginine
	CUA: Leucine	CCA: Proline	CAA: Glutamine	CGA: Arginine
	CUG: Leucine	CCG: Proline	CAG: Glutamine	CGG: Arginine
A	AUU: Isoleucine	ACU: Threonine	AAU: Asparagine	AGU: Serine
	AUC: Isoleucine	ACC: Threonine	AAC: Asparagine	AGC: Serine
	AUA: Isoleucine	ACA: Threonine	AAA: Lysine	AGA: Arginine
	AUG: Methionine	ACG: Threonine	AAG: Lysine	AGG: Arginine
G	GUU: Valine	GCU: Alanine	GAU: Aspartic acid	GGU: Glycine
	GUC: Valine	GCC: Alanine	GAC: Aspartic acid	GGC: Glycine
	GUA: Valine	GCA: Alanine	GAA: Glutamic acid	GGA: Glycine
	GUG: Valine	GCG: Alanine	GAG: Glutamic acid	GGG: Glycine

but no ambiguity. For example, AGA and AGG both specify Arginine, but neither identifies another amino acid. It is notable that all life (plant or animal) on Earth uses the same genetic code. This suggests that while there may have been numerous unrelated occurrences of life originating on the primordial Earth, it appears we are all descendants of one common primordial ancestor.

If you were to take an RNA molecule, make a mirror image of it and put the two together, you get DNA (see Figure 6.9). The mirror image of the base adenine (A) is thymine (T) and of guanine (G) is cytosine (C). Together A and G are referred to as purines and C and T as pyrimidines. Within the nucleus of each of your cells lies a DNA molecule. If you were to unwrap a human DNA you would find it to be ~1 meter long and contain ~3 billion base sequences! These base sequences can be grouped into ~1 billion codons, which reside in 23 pairs of chromosomes during cellular reproduction/division. Each chromosome in turn contains hundreds to thousands of genes (see Figure 6.10). In comparison, a simple bacterium contains a sum total of ~1,000 genes.

6.2.3 Lipids

A lipid is a biologically produced, largely nonpolar substance composed of hydrocarbon chains which do not make bonds easily with water (a polar molecule). This renders them insoluble in water. Examples of lipids include oils, fats, fatty acids, and cholesterol. Being insoluble in water makes lipids a viable material from which cellular membranes and smaller specialized structures within a cell (i.e., organelles) can be formed. A cell membrane is a type of lipid bilayer that can be formed through the hydrophobic effect. One type of lipid called a phospholipid is polarized on one end (the head) and unpolarized on the other (the tail). When surrounded by water, the polar heads of these lipids align toward the

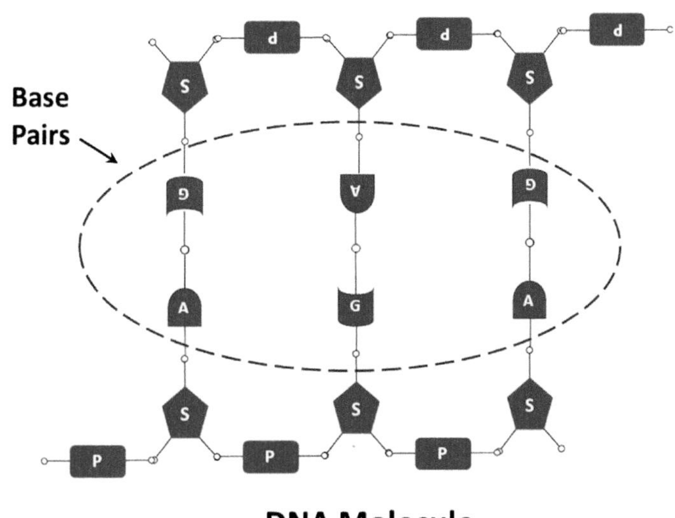

FIGURE 6.9 DNA structure. By taking the mirror image of the RNA molecule of Figure 7.8 and attaching the two, a DNA molecule is produced. Here a three base pair sequence designating Arginine is shown. The mirror image of the base A is G. By knowing one half of a base pair, it is possible to deduce the other.

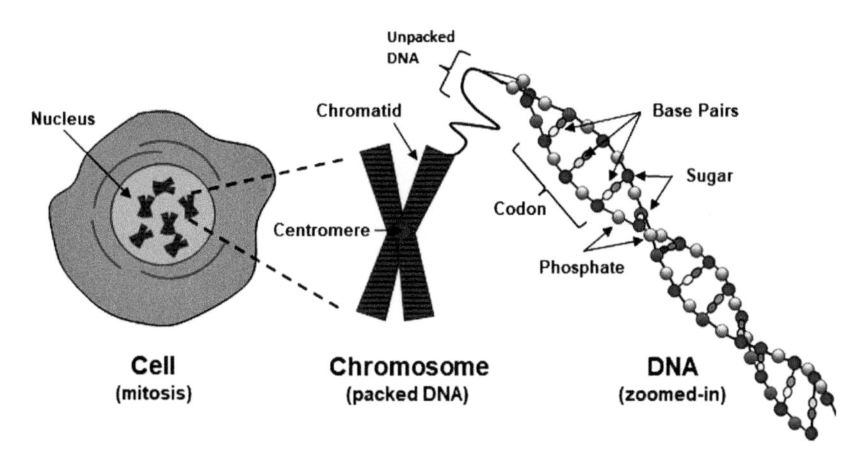

FIGURE 6.10 Relationship between base pairs, genes, and chromosomes. A gene is a segment of a DNA molecule containing several hundred codons. During cell division, i.e., mitosis, the DNA is coiled up into a chromosome structure. About 95% of the time the DNA is loose/unpacked within the nucleus to make the base pairs more accessible to messenger RNA. The cell cycle, the series of events that take place in a cell as it grows and divides, is (for mammalian species) ~24 hours.

water molecules, while the nonpolar tails orient themselves away from water. Depending on the concentration of the lipid, this interaction can lead to the formation of flat membranes or spherical structures (see Figure 6.11). These structures could potentially serve as the protective vessels for more complicated molecules, such as nucleic acids. The formation of lipids into protocell membranes could be a key step in the origin of life.

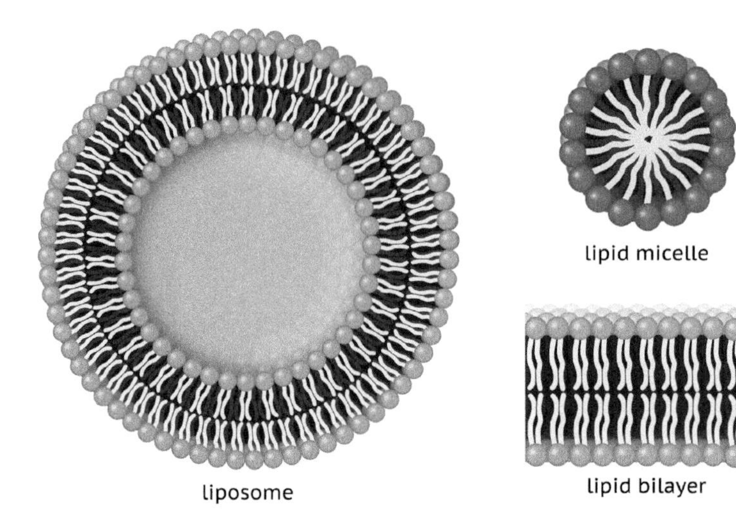

FIGURE 6.11 Self-organization of lipids. We all know that oil (composed of lipids) and water do not mix. Indeed, some parts of a lipid molecule are so hydrophobic that they turn their backs to water molecules and, in the process, create either spherical shell or pancake-like structures. Such structures could have served as the first protocells, within which nucleic acids could evolve and spark the existence of life. Image: Shutterstock.

6.2.4 Carbohydrates

When one hears the term carbohydrates, images of starchy foods such as cereal, bread, potatoes, and pasta comes to mind. But in biochemistry the term refers to a family of biogenic molecules that also includes sugars, phosphates, and cellulose. All three types of carbohydrates are essential to the existence of life. Sugars, in partnership with phosphates, form the structural backbone of RNA and DNA, as well as serve as an energy source to drive metabolism. Plants use cellulose as a primary structural material. Carbohydrates have the chemical formula $C_m (H_2O)_n$ (where m and n are integers). An example of this structure can be seen in the schematic of the ribose sugar shown in Figure 6.6. Some carbohydrates, like glucose, can assume a chain or ring geometry. The ubiquity of H_2O in biogenic molecules like carbohydrates and its role in facilitating chemical reactions underscore its importance to the existence of life on Earth.

6.3 DOMAINS OF LIFE

In 1665 the English scientist Robert Hooke examined the appearance of cork under a primitive microscope. He observed many box-like structures and described them as being similar in appearance to the cells inhabited by monks in a monastery (see Figure 6.12). The name stuck.

As the atom is for elements and the photon for light, the cell is for all known life. A cell is the smallest structural unit within which all the functions required for life are contained. Unlike a virus, a cell has all the machinery to replicate itself. Indeed, there exists a host of life forms (e.g., bacteria and protozoa) that are unicellular. More advanced forms of life (e.g., plants and animals) are multicellular. Each human is an assembly of ~30 trillion cells.

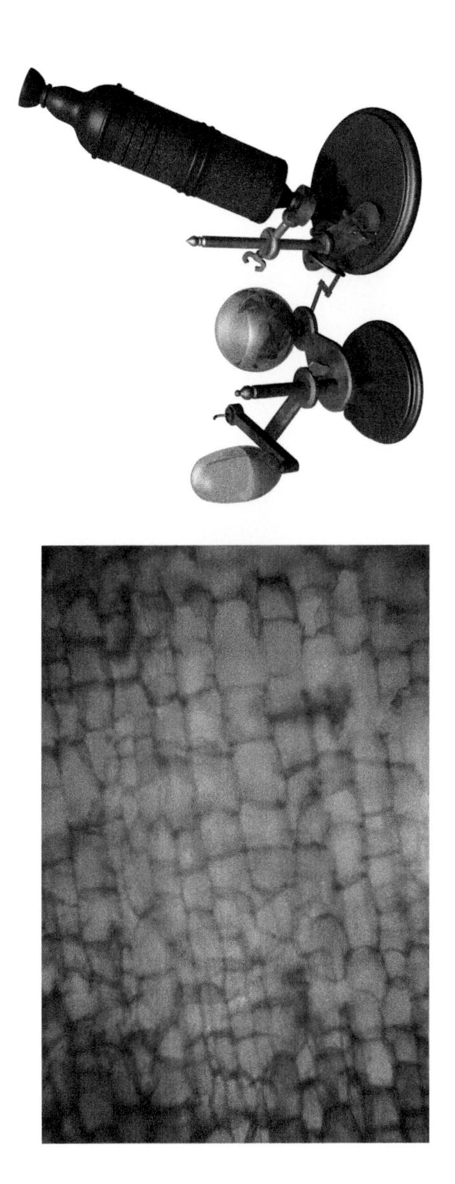

FIGURE 6.12 Appearance of cork cells in microscope (Left). A microscope designed and used by Robert Hooke (Right). Images: Shutterstock.

About 330 billion of them are replaced daily (Fischetti and Christiansen 2021), a value approximately equal to the number of stars in the Milky Way.

The two main types of cells found on Earth are eukaryotes and prokaryotes. Schematic representations of each are shown in Figures 6.13 and 6.14. Table 6.2 lists their similarities and differences. Plants, animals, fungi, and protozoa are mostly made of eukaryotic cells. Such cells are complex, with a distinct nucleus containing DNA arranged into chromosomes. Within the cell body are membrane-bound structures called organelles. There are different types of organelles, each with a specific function (see Table 6.3). Simpler lifeforms such as bacteria and archaea have a prokaryotic structure, with no distinct nucleus or

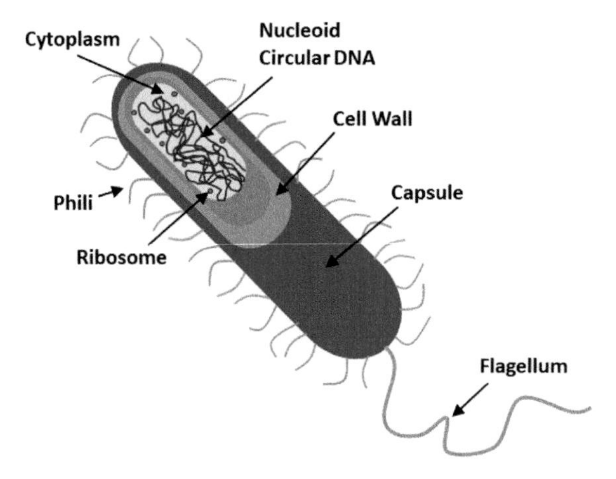

FIGURE 6.13 Prokaryotic cell structure. The genetic material (circular RNA or DNA) is contained in an irregularly shaped region called the nucleoid, with no nuclear membrane. The Phili are hair-like structures that make the cell stick to other host cells (like spikes or hooks on a burr). Movement of the flagellum allows the cell to swim.

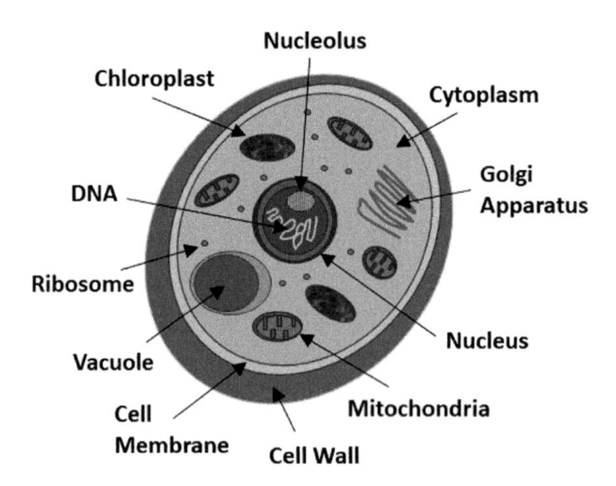

FIGURE 6.14 Eukaryotic plant cell structure. Eukaryotic cells have a well-defined nucleus and utilize helical DNA. They are mostly round or spherical and contain a number of organelles, each with a specific function; these include mitochondria and chloroplasts (energy sources) and ribosomes (protein construction sites).

TABLE 6.2 Cell Type Comparison

	Eukaryotic	Prokaryotic
Nucleus	Yes	No
Genetic Material	DNA	DNA
Cell Wall	Depends	Always
Cell Membrane	Yes	Yes
Membrane Organelles	Yes	No
Ribosomes	Yes	Yes
Size	10–100 μm	0.1–10 μm

TABLE 6.3 Organelle Functions

Nucleus	Contains Nucleic Material
Mitochondria	Major energy source in plant and animal cells through production of ATP; has its own DNA
Chloroplast	Contains chlorophyll, produces energy reserves in plants in the form of ATP and NADPH, and in the process, freeing oxygen from water; has its own DNA
Ribosome	Site for biological protein synthesis; has its own RNA
Lysosome	Digests cellular waste products; found in animals
Vacuole	Contains cell sap or other material (large vesicles)
Vesicles	Hollow spheres used to transport substances
Golgi Apparatus	Packages proteins in vesicles for transport (post office)

organelles. Prokaryotes have a single, free-floating, circular DNA molecule. Eukaryotes are mostly round or spherical, while prokaryotes can exist in the form of rods (bacillus), spheres (cocci), curves (vibrio), or even spirals (spirochetes). The relative simplicity of prokaryotes suggests they were the first lifeforms on Earth.

Archaea (see Figure 6.15) were once considered a form of bacteria and are similar in size and shape but have genes and chemical reactions (metabolic pathways) more similar to those in eukaryotes. This realization has led to the idea that the classification of life

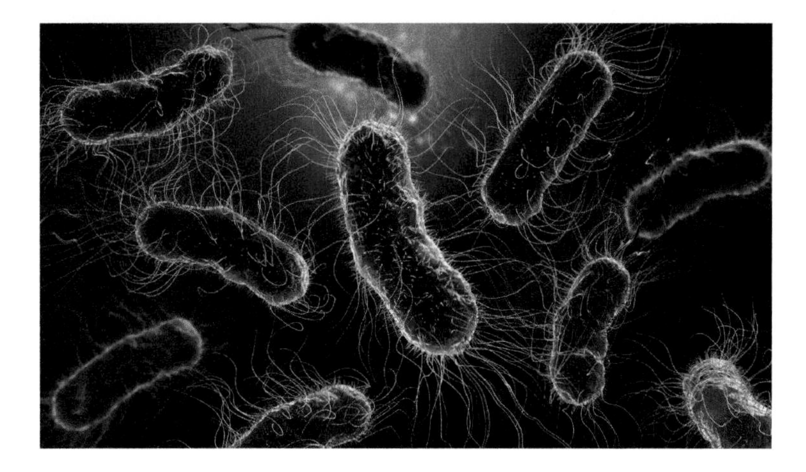

FIGURE 6.15 Example Archaean. Halobacterium thrives in warm, salty environments such as the Great Salt Lake and the Dead Sea. Each cell is about 5 microns long. Image Credit: Shutterstock.

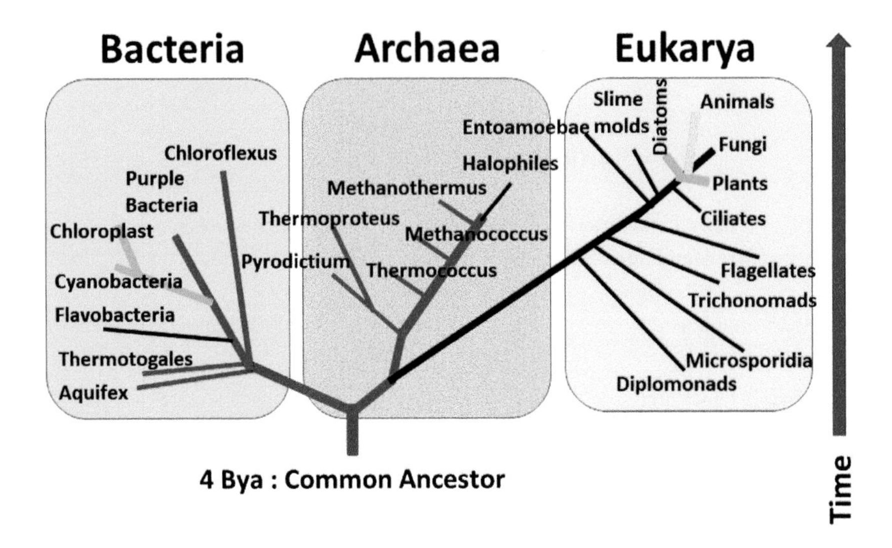

FIGURE 6.16 Phylogenetic tree of life: the three domains of life are bacteria, archaea, and eukaryotes. Lifeforms in the bacteria and archaea domains are both often referred to as Prokaryotes due to their lack of a well-defined nucleus. All three domains are believed to have originated from a common ancestor within ~600 million years of the Earth's formation. Viruses may form yet another domain of life. Red branches trace the evolution of hyperthermophiles that require a minimum growth temperature of 80°C. Green branches trace oxygenic and blue-branched anoxygenic (i.e., oxygen and nonoxygen producing) photosynthetic organisms. Unlike the more familiar photosynthesis in plants where water, light, and CO_2 yield energy and oxygen, in anoxygenic photosynthesis the combination of hydrogen sulfide, light, and CO_2 yield energy and sulfur. The yellow limb of the tree corresponds to animals. Adapted from Farmer 2013.

as either prokaryotic or eukaryotic should be replaced with a three-domain classification as archaea, bacteria, or eukarya (Woese, Kandler, and Wheelis 1990). Figure 6.16 is an evolutionary tree, also called a phylogenetic tree, that attempts to illustrate the evolutionary relationships between species based upon similarities and differences between their physical or genetic characteristics. Viruses, if considered alive, would belong in a fourth domain (Nasir and Caetano-Anolles 2015). Archaea utilize enzymes during replication similar to those used by eukaryotes. Unlike any other cells, archaea use ether lipids in their cell membranes and can utilize a wide range of unique energy sources, including ammonia, metal ions, and hydrogen gas – potentially caustic to other lifeforms. These properties make archaea extremophiles capable of thriving in extreme environments (e.g., volcanic vents, hot springs, and salt lakes), places common on a primordial Earth or even today within the moons of the outer planets. More recently, they have been found to exist in much more mundane locations, such as the human oral cavity, skin, and colon. Fossil remains of archaea cells have been found in 3.5 billion-year-old stromatolites – mounds of fossilized cyanobacteria (Figure 6.17). Which came first, archaea or bacteria, is still a subject for debate, but their adaptability, relative simplicity, and ubiquity make archaea a contender for one of the oldest lifeforms on Earth. Life on Earth has had over 3.5 billion years to become finely tuned to its surrounding. The simplicity and diversity of archaea make their study important for preparing for what we might encounter in other worlds.

FIGURE 6.17 Stromatolites. Located in Hamelin Bay Pool Marine Nature Reserve, Shark Bay, Western Australia, stromatolites are the fossilized remains of some of the oldest known life-forms. The stromatolites shown above are believed to have formed ~3.45 billion years ago. Image: Shutterstock.

Figure 6.16 would seem to suggest that as life evolved it passed through one or two domains before reaching its final destination, but this is not necessarily the case. The two organelles within eukaryotic cells, mitochondria and chloroplasts, contain their own genetic material and suggest that over 1 billion years ago they may have evolved and lived independently within the primeval sea as prokaryotes. Then, as chance would have it, one day migrated into larger eukaryotic cells and entered a symbiotic relationship (Margulis and Sagan 1986). Their presence provided enough energy not only to keep themselves going but to also "supercharge" (literally) the host cell's metabolism. The increased metabolism may have provided the necessary kick to drive increased reproduction and, thereby, evolution to the point where more complex multicellular organisms could come into existence.

Cells that incorporated chloroplasts became plant cells, which through photosynthesis could combine light, water, and carbon dioxide to produce energy for the cell and release molecular oxygen into the atmosphere. Indeed, this build-up of atmospheric oxygen had been going on since the evolution of prokaryotic cyanobacteria ~3.5 billion years ago. Cyanobacteria from this epoch combined with sedimentary grains in shallow water to form microbial mats, leading to the formation of stromatolites (see Figure 6.17). It is from cyanobacteria that chloroplasts are believed to have originated (Kumar, Mella-Herra, and Golden 2010). This seems like a good thing, but at this point, all life on Earth had evolved in an oxygen-free environment, so did not require it. They were anaerobic. In fact, oxygen itself is a very corrosive gas and can even dissolve metal (by rusting) over time. Eventually the build-up of atmospheric oxygen emitted by cyanobacteria destroyed countless species of anaerobic lifeforms. This was one of several times of great dying on the Earth and is sometimes referred to as the "Oxygen Crisis" estimated to have peaked about 2.45 billion years ago (Torres et al. 2015). Fortunately, the other energy-producing organelle, mitochondria, also once believed to be an independent prokaryotic organism, had evolved the ability to utilize oxygen to efficiently extract energy from ATP molecules. Approximately 2 billion years ago a primordial prokaryotic mitochondria-like organism invaded a eukaryotic cell and entered into a symbiotic relationship; the larger eukaryote cell provided a sheltered environment for the mitochondria and, in exchange, the mitochondria released enough energy to boost the metabolism in the host cell by an order of magnitude. What was once a poisonous gas, produced as an unfortunate byproduct of photosynthesis, was used to propel evolution forward. Cells that utilize oxygen are referred to as aerobic. Every cell in your body contains the descendants of the humble, primordial mitochondria. If it were not for this chance encounter between a mitochondria-like prokaryote and a eukaryotic cell in a murky pool of water between 1.7 and 2 billion years ago, we would not be here (Emelyanov 2001; Feng, Cho, and Doolittle 1997).

6.4 CELL FUNCTION

Water makes up ~75–85% of the cell's mass and volume. As mentioned above, the walls of a cell are made of phospholipids which serve as a "castle wall" that isolates the content of the cell from the environment in which it finds itself. But like all castle walls there are gates through which "goods and service" can pass. In the case of the cell the role of gatekeeper and doorway is played by certain protein molecules that stud the cell wall and determine which substances have the proper chemical passwords to enter or leave. The cell contains cytoplasm and a nucleus. The cytoplasm consists of a jelly-like substance called cytosol within which can be found the organelles of Table 6.3. Functionally a cell can be thought of as a miniature, self-contained city, with borders set by the cell wall, chloroplasts, and/or mitochondria acting as energy producers and consumers, distribution centers composed of Golgi bodies, vesicle serving as supply/garbage trucks, and a city hall nucleus dispensing directives (Figure 6.18).

The nuclear DNA not only contains all the information required to reproduce a cell, it also controls the timing and nature of metabolic activities within the cell, such as the production of proteins. Figure 6.19 illustrates the process by which DNA instructs the cell to create a protein. When the DNA determines it is time to make a protein the part of the DNA ladder containing the codon or codons required to specify the protein splits open, exposing the corresponding base sequences. Phosphates, sugars, and bases floating within the nucleus then self-assemble into a complementary length of RNA that then detaches itself and migrates from the nucleus into the cell body. Since the RNA is conveying a message from the nucleus, it is referred to as messenger RNA or mRNA. The mRNA proceeds on its journey until it is chemically captured by a ribosome. Ribosomes are themselves made from a 60/40 mixture of ribosomal RNA, rRNA, and proteins. Also floating around in the cell is transfer RNA or tRNA. Each tRNA is one-codon long and carries on its back the amino acid corresponding to its codon base sequence. While the ribosome holds the

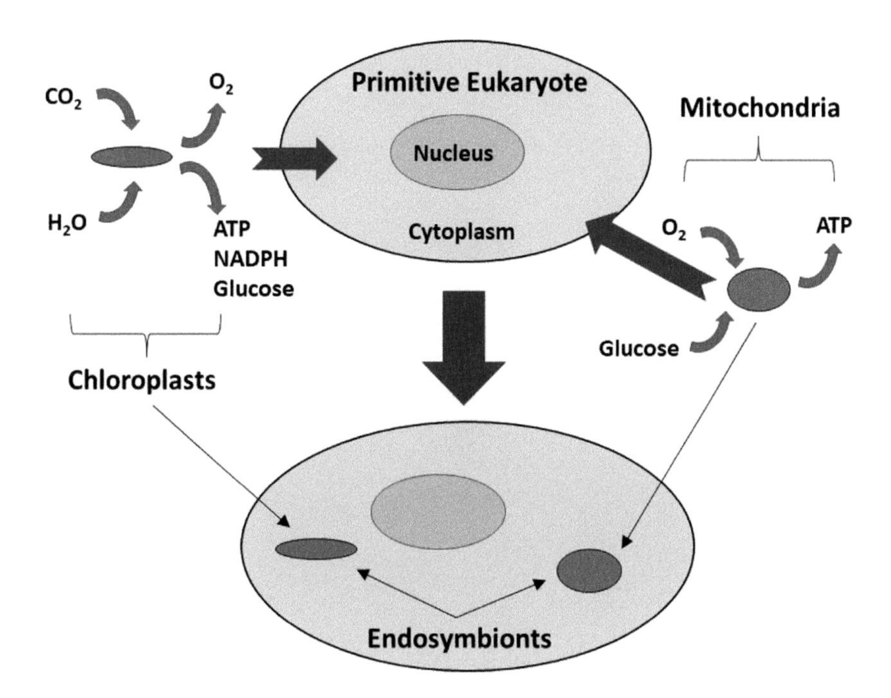

FIGURE 6.18 The primordial atmosphere of Earth came from the outgassing of volcanoes and consisted mostly of nitrogen (N_2), carbon dioxide (CO_2), and water (H_2O). It did not contain appreciable amounts of oxygen (O_2). At this point in time all life on Earth was anaerobic. Then, ~3.5 billion years ago, a prokaryotic cell (cyanobacteria) evolved the ability to use photosynthesis to combine sunlight, CO_2, and H_2O to generate energy (in the form of glucose, i.e., sugar). For the existing lifeforms, an unfortunate byproduct of this process was molecular oxygen, a highly corrosive gas. As oxygen built-up in the Earth's atmosphere more and more lifeforms died. This time in the Earth's history is often referred to as the "Oxygen Crisis". To exacerbate matters, the oxygen-producing prokaryote (an ancestor to the modern cellular chloroplast) entered into an endosymbiotic relationship with a more evolved eukaryotic cell. The added energy provided by photosynthesis dramatically increased the reproduction rate of the host eukaryotic cell. Over time, this led to a further increase in the level of O_2 in the atmosphere. Fortunately, 1.7–2 billion years ago another prokaryotic organism evolved the ability to utilize O_2 to increase its ability to break down glucose and release energy, thereby increasing its ability to survive and reproduce. Eventually this prokaryotic organism also entered into an endosymbiotic relationship with a eukaryotic cell, instilling it with supercharged reproductive powers, thereby accelerating the evolution of life on Earth.

mRNA in place, multiple tRNAs attempt to lock on to it. Only those tRNAs with complementary codons are able to chemically hook up. As the amino and carboxyl groups of the "piggy back" amino acids come into close proximity, a peptide bond forms securing them together. The mRNA moves through a ribosome like a bicycle chain on a chemical sprocket. Once the complete mRNA chain is read and the specified protein synthesized, the tRNAs break off from the protein and are recycled into the cell to pick up their next amino acid load.

How long this process takes varies from organism to organism and depends on the length of the protein being synthesized. The transcription of three base pairs (one codon)

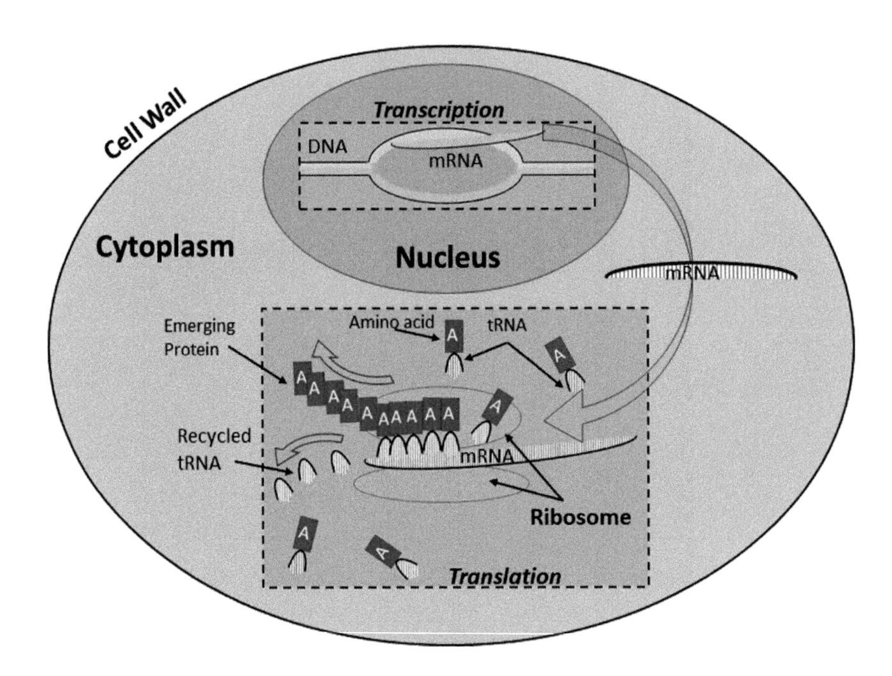

FIGURE 6.19 Protein synthesis in the cell. When a particular protein is needed within the cell the nuclear DNA ladder opens up and exposes the base sequence that identifies the protein. The required base sequence is then transcribed into mRNA that self-assembles about the exposed base sequence. The mRNA then migrates from the nucleus into the cytoplasm where it is captured by a ribosome. The ribosome is where the mRNA is translated into a protein. A ribosome is composed of two subunits. The smaller subunit can be thought of as a chemical "sprocket" over which the mRNA "chain" passes. tRNA is also present in the cytoplasm. Each tRNA carries on its back an amino acid corresponding to its base sequence. The larger ribosome unit is the site where the various tRNAs attempt to "hook-up" with a codon of the mRNA. When the proper tRNAs are aligned side-by-side their amino and carboxyl groups form a peptide bond, welding the amino acids together. ATP is used to generate the heat required to form the bond. ATP is created using electrical energy produced through chemical reactions in mitochondria. (One byproduct of ATP production is body heat.) The mRNA "chain" will continue moving over the ribosome "sprocket" until the full protein is formed. As the newly formed protein emerges from the ribosome the tRNA carriers are recycled and pick up another amino acid load to be used in the next protein synthesis.

from the nuclear DNA to mRNA takes about 0.3 sec, while the translation into the corresponding amino acid by a ribosome takes about half as long. For example, if a protein requires 1,200 base pairs (or nucleotides) to describe it, about 2 minutes will be required to transcribe it into mRNA and ~20 sec to translate it into a 400 amino acid long protein. Because prokaryotes (e.g., bacteria) are simpler, the whole process can happen about twice as fast.

6.5 ORIGIN OF LIFE

Debates concerning the possibility of life emerging spontaneously from inorganic matter have been documented since the time of the ancient Greeks. Both Anaximander (610–546 BC) and Aristotle (384–322 BC; Aristotle 1912) discussed it at length. Up until the time

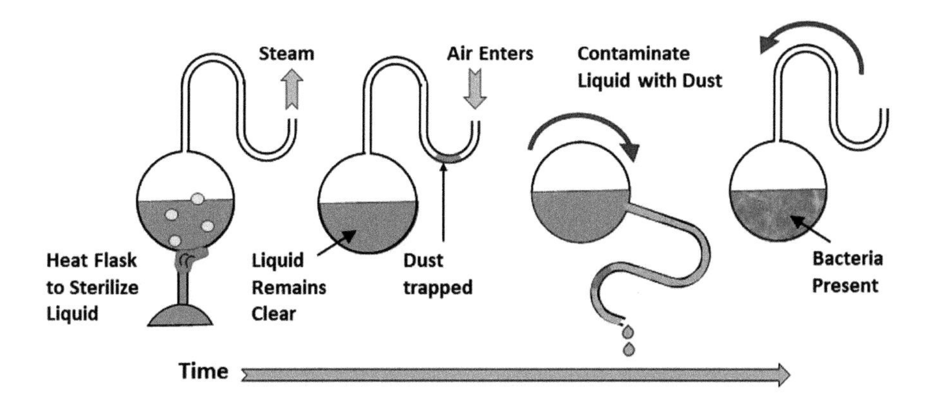

FIGURE 6.20 Spontaneous Generation Experiment (1859). In an experiment to disprove the long-held theory of spontaneous generation, Louis Pasteur first sterilized liquid by boiling it in a swan-neck flask. Steam escaped through the opening in the neck. He then removed the flame and let the flask sit for several days. During that time dust carrying bacteria entered the test set-up but got trapped in the curve of the neck. If spontaneous generation was correct, then bacteria would start growing in the liquid. However, the liquid remained clear. It was only after Pasteur briefly tipped the flask over so that the liquid came in contact with the trapped dust, that bacteria started growing in the flask.

of Louis Pasteur, it was a commonly held belief that such lifeforms as fleas could arise spontaneously from dust and maggots from rotting flesh. In 1859 Pasteur disproved this belief through an elegant experiment where he boiled a meat broth in a special curved, "swan-necked", flask designed to let air in, but keep microorganisms out (see Figure 6.20). The microorganisms in the air (if they existed) would settle by gravity into the deep curve of the flask's neck before reaching the broth. After boiling he showed there was no microorganism growth in the broth unless he tilted the flask so the broth could reach the lowest point in the neck where microorganisms were trapped. In this one experiment, he was able to show both that spontaneous generation, at least in the classical sense, did not occur and that microorganisms are in the air all around us (Ball 2016).

So, the question remains, how did life evolve from inorganic matter? Sixty-five years after Pasteur's experiment the Russian biochemist Alexander Oparin and British biologist J. B. S. Haldane independently proposed life could indeed originate from inorganic material, but only through a slow evolution from a "primordial soup" in a reducing environment. They theorized such an environment is needed to facilitate the formation of the hydrogen-rich compounds required for life. Shortly after this new pathway to biogenic molecules was proposed both ammonia (NH_3) and methane (CH_4) were spectroscopically discovered in the atmosphere of Jupiter. This discovery lent support to the concept of the Earth having a primordial reducing atmosphere and Oparin's and Haldane's theory.

6.5.1 Miller–Urey Experiment

In a laboratory at the University of Chicago in 1952 Stanley Miller and Harold Urey put these ideas about the origin of life on primeval Earth to the test in their now famous Miller–Urey experiment (Miller 1953). Their experimental design is illustrated in Figure 6.21.

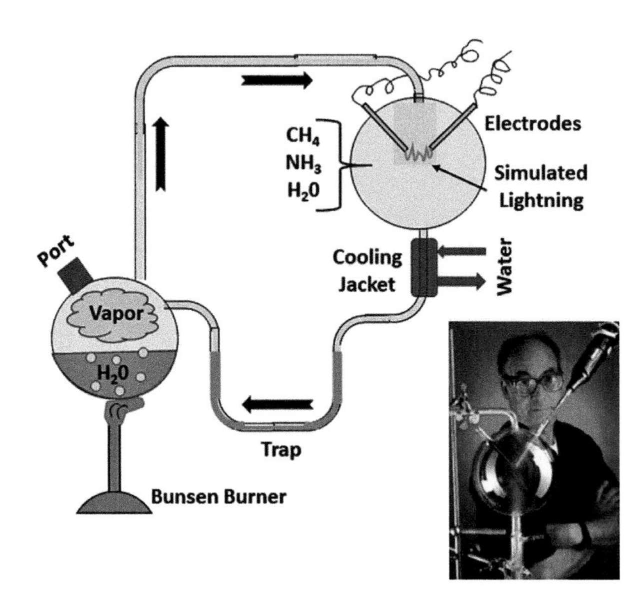

FIGURE 6.21 Miller–Urey Experiment. In 1953, Stanley Miller and Harald Urey designed and operated an experimental apparatus with the goal of simulating conditions on the early Earth. Within a day of operation, pink organic material appeared in the experiment's trap. The experiment demonstrated for the first time that amino acids, the building blocks of life, could have been formed naturally on (or in) the primordial Earth. Inset: Stanley Miller at University of Chicago. Image Credit: Alamy.

They assembled glass flasks, spheres, and tubes into a closed-loop system to simulate the oceans, atmosphere, and energy sources present on an early Earth. The whole apparatus was first sterilized and then filled with gaseous methane, ammonia, and hydrogen (primordial atmosphere). One flask contained water (early oceans) that was heated by a flame (geothermal energy). Water vapor together with the other gases circulated through the apparatus encountering first a sphere with simulated lightning bolts going through it and then a cooling jacket to make it rain. After letting the experiment run for one day, the solution gathered in the trap turned pink in color. After a week of operating, the flasks and tubes became covered in goo. Examination of the goo showed it to be composed of a wide variety of complex organic molecules, including over 20 amino acids. No lifeforms crawled out of the flasks, but the experiment did successfully show how the organic building blocks of life could be manufactured quite easily and quickly from prebiotic compounds.

Subsequent studies of the Earth's early atmosphere have shown it to be less reducing than that simulated by Miller and Urey (Green 2011). However, as long as there was no O_2 (that would not come for another ~2 billion years), similar organic compounds as those found by Miller and Urey are created (Green 2011). Subsequent experiments have shown that adding hydrogen cyanide (HCN) to the mix can lead to the production of large amounts of adenine ($H_5C_5N_5$), one of the purine bases used in RNA and DNA (Oró and Kimball. 1961). Other organic molecules necessary for life can be formed from raw materials present on early Earth. For example, the pyrimidine base cytosine can be made by combining HC_3N and urea. Ribose sugar can be produced from H_2CO and heat, while

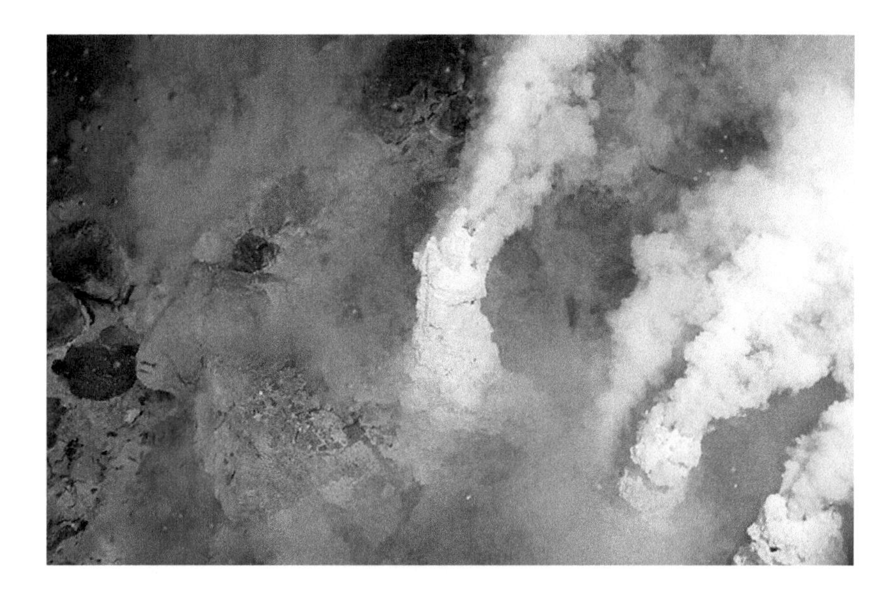

FIGURE 6.22 White smokers emitting 103°C liquid carbon dioxide bubbles from a hydrothermal vent in the Champagne vent site of the NW Eifuku volcano. The white chimney are ~20 cm across and ~ 50 cm high. Image: Photo courtesy of NOAA.

the other structural element required for RNA/DNA, phosphates, is a simple byproduct of rock erosion.

One place on Earth where many of the conditions of the Miller–Urey experiment can be met is in the vicinity of seafloor hydrothermal vents. Here there is water in abundance, hydrogen-rich compounds, a heat source, and no oxygen (see Figure 6.22). Seafloor vents occur where water is geothermally heated and flows through a fissure in the ocean's crust. At these locations, they may form either black or white smokers, depending on mineral content. Compared to other places on the sea floor hydrothermal vents are biologically active, with chemosynthetic bacteria and archaea forming the base of the food chain. Here can be found unworldly giant tube worms that grow as much as 2.4 meters long, as well as more familiar species such as shrimp, snails, and clams. Hydrothermal vents have been observed spewing material into space from the surface of Jupiter's moon Europa and Saturn's moon Enceladus (Figure 6.23). Such vents will likely serve as prime locations for searching for life elsewhere in the solar system and beyond. Given the highly volcanic nature of the early Earth, seafloor hydrothermal vents must have been common place, pumping many tons of prebiotic raw materials into the primordial seas each day.

The closest terrestrial surrogate to Europa's subterranean ocean is Lake Vostok, located 1,417 ft below the Russian Vostok Base in Antarctica (Figure 6.24). Established in 1957, the location of the base was chosen due to its proximity to the Earth's southern magnetic pole and the deep ice core drilling possibilities it offers. By analyzing air bubble trapped in the ice at different depths, it is possible to determine the composition of the Earth's atmosphere at the time the ice formed, thereby informing paleoclimate models. Vostok is one of the coldest places on Earth, having reached −89.2 °C (−128.6 °F) on July 21, 1983 (Budretsky 1984). Airborne ice-penetrating radar survey of the region in the early 1970s suggested the

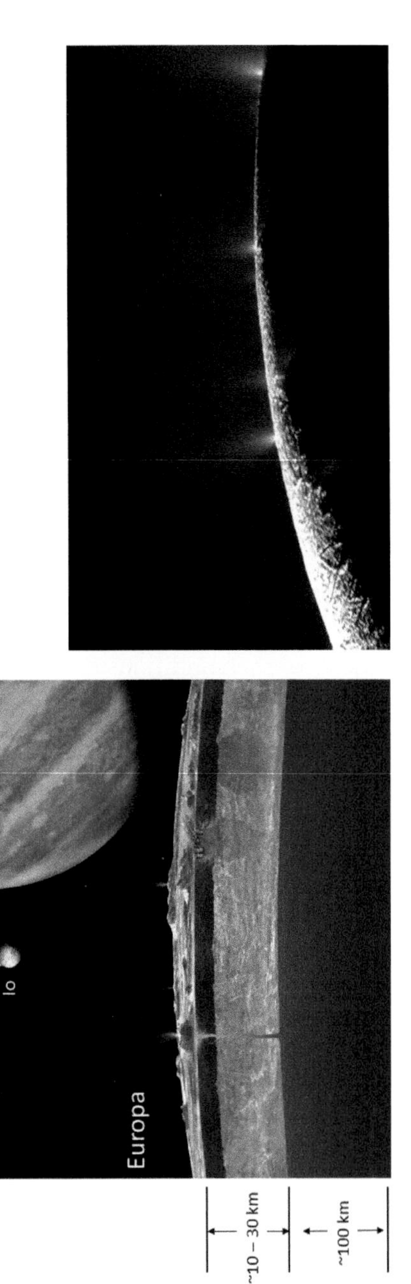

FIGURE 6.23 Off-world hydrothermal vents. (*Left*) Artist conception of plumes observed by the Hubble Space Telescope emerging from Europa's subterranean sea. (*Right*) Cassini flyby mission photograph of water vapor venting from Saturn's Moon Enceladus. The existence and frequency of such vents within our solar systems suggests they may be a common occurrence in planetary systems. Image Credit: NASA/JPL.

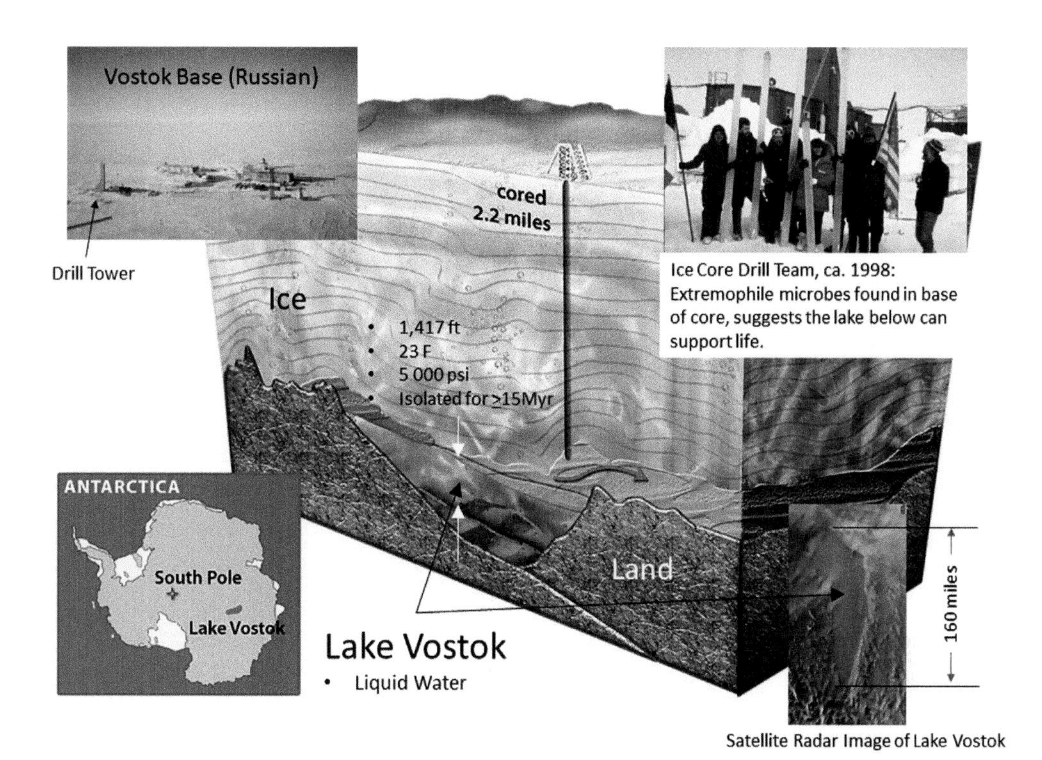

FIGURE 6.24 Lake Vostok. Located 1,417 ft below the Antarctic ice sheet, Lake Vostok may contain lifeforms that have been isolated from the Earth's surface for millions of years. Campaigns and methodologies for studying Lake Vostok serve as valuable "dress-rehearsals" for future missions to search for life below the icy surfaces of worlds in the outer solar system. Image Credits: NSF/NASA/NOAA.

presence of a fresh water lake beneath the ice (Oswald 1973). Further radar observations from space and air later confirmed the presence of Lake Vostok (Kapitsa et al. 1996). The average water temperature is estimated to be 27 °F (–3°C). The pressure of the overlying ice layer keeps it from freezing. The insulating effects of the ice layer and geothermal heating from below may also play a role in maintaining a liquid state. The lake has been sealed off from the atmosphere since its formation over 15 million years ago. The analysis of ice cores taken in 1998 provided the first evidence of extremophile microbes within 100 m of the top of the lake (d'Elia et al. 2009). Subsequent studies of ice samples taken both above and from the lake itself suggest the presence of bacteria, eukaryotes, and perhaps even fish in the lake (Gura 2020). However, the possibility of sample contamination by the associated drilling equipment is a concern (Schiermeier 2013).

The experience gained from studying Lake Vostok will be invaluable in the design of future sample-return missions from icy worlds that could harbor life in the outer solar system.

6.5.2 Chemical Evolution to Life

The chemical evolution to life can be divided into three steps:

- the synthesis of monomers

- the synthesis of monomers to polymers
- the transition to life

The Miller–Urey experiment showed that the building blocks of life, amino acids (monomers), were likely common on early Earth. How likely is it that monomers would combine to form polymers? As discussed in Section 6.2 polymers are formed from "train cars" of monomers by way of an endothermic peptide bond. Water plays a critical role in both the creation and destruction of peptide bonds, as well as serving as the medium through which monomers come into contact with one another. Without water, life literally shuts down. This would suggest the synthesis of monomers to polymers likely occurred in a place where water is present, but not all the time, and where there is heat to activate peptide bonds. One such place that comes to mind is a tidal pool, like that shown in Figure 6.17. Here water-carrying amino acids, perhaps generated in nearby seafloor vents, washes in and out of tidal pools each day. While in such pools light from the Sun could warm the pool providing the heat source for activating peptide bonds. As far back as 1924, Oparin showed that organic molecules in solution may spontaneously form droplets and layers that could potentially serve as protocells (see Figure 6.11). Each of the nearly countless protocells that then existed on Earth could potentially serve as a small laboratory within which chemical evolution could continue. Within the protective environment of a protocell delicate, complex chains of molecules could form, eventually yielding the first RNA. A protocell with active RNA could be classified as a prokaryote and the first lifeform on Earth. To enhance the likelihood of survival, evolution typically pushes lifeforms to ever greater complexity; from prokaryotes to eukaryotes and, ultimately, multicellular lifeforms such as us.

How nucleic acids might have come to exist within a protocell and begin working so closely with proteins? One idea is that eventually a protein was formed in a primordial protocell that could polymerize nucleotides – actually string them together – into the first RNA. Such proteins do exist. RNA replicase is an enzyme (specialized protein) found in modern RNA-containing viruses that aid in the copying of RNA from template or "seed" RNA (Koonin, Gorbalenya, and Chumakov 1989). Even without seed RNA, replicase will attempt to "create" RNA out of a mixture of phosphates, sugars, and bases. RNA replicase as it is found today probably did not exist on early Earth, but something similar might have (Evans 1996). Efforts to create artificial cells are ongoing. Recent advances in microfluidic technologies are accelerating work in this important field (Pelletier, et al. 2021; Powell 2018; Litschel, et al. 2018).

SUMMARY

What made the primordial Earth conducive to the origin of life would render it inhospitable to many of the lifeforms we know today. The surface conditions on the primordial Earth, for example, volcanic, a mildly reducing atmosphere, oceans, sea floor vents, storms, tidal forces, cometary impacts, etc., are things that will also have been present on many, many terrestrial worlds orbiting Sun-like stars. The fact that life originated so early in Earth's history suggests that, under similar conditions, it will happen again. The same laws of nature that governed the origin and evolution of life here will be at work there. Or, perhaps, we were just lucky ~4 billion years ago and life is a rare occurrence. The history of astronomy informs us that every time humanity has thought it held a privileged position in the cosmos, we later found ourselves to be wrong. In the coming decades, we will continue our push outward into the solar system. We will carefully sift through the environs of neighboring worlds for evidence of life. If we do find evidence of indigenous life in any of our sister worlds, then the age-old question of whether life occurred elsewhere will be answered. Indeed, such a find would suggest the galaxy is teaming with life. The answer will likely be revealed within our lifetimes. In this regard, we are indeed privileged.

REVIEW QUESTIONS

1. What were conditions like on the Earth soon after it formed?
2. Compare these conditions to those of the Urey–Miller experiment.
3. Where on Earth and in the solar system are conditions similar to those of the Urey–Miller Experiment?
4. What are the four molecules of life?
5. What are their functions?
6. What is the length of a human DNA molecule?
7. What are DNA molecules made of?
8. What is a peptide bond?
9. What is a ribosome?
10. What is the simplest amino acid?
11. What makes lipids useful as cellular membranes?
12. What is the difference between tRNA and mRNA?
13. How is information stored in a DNA molecule?
14. How is the information contained in DNA conveyed to the cell? How long does this process take?
15. What is a prokaryote and eukaryote?
16. What are some examples of organelles?
17. How does DNA copy itself?
18. What is the Phylogenetic Tree of Life? Where are we on it?
19. About how many cells are in the human body?
20. About how long does it take for a cell to replace itself?

21. How long does it take to replace all the cells in your body?
22. What evidence is there that all life on Earth has a common ancestor?
23. What are the two types of bases?
24. What is a codon?
25. What is a chromosome?
26. What is an anaerobic lifeform?
27. What was the Oxygen Catastrophe?
28. How did life adapt to the presence of oxygen?
29. What evidence is there that the spontaneous generation of life (at least in the short term) does not happen?
30. What is the difference between aerobic and anaerobic life?
31. Why is Lake Vostok important?

REFERENCES

Aristotle. 1912 [c. 350 BCE]. "Book III." *On the Generation of Animals*, translated by Arthur Platt. Oxford: Clarendon Press. ISBN 90-04-09603-5. Accessed September 1, 2009.

Budretsky, A. B. 1984. "New Absolute Minimum of Air Temperature." *Bulletin of the Soviet Antarctic Expedition (in Russian).* Leningrad: Gidrometeoizdat, p. 105.

d'Elia, T., R. Veerapaneni, V. Theraisnathan, and S. O. Rogers. 2009. "Isolation of Fungi from Lake Vostok Accretion Ice." *Mycologia* 101 (6): 751–63.

Emelyanov, V. V. 2001. "Rickettsiaceae, Rickettsia-Like Endosymbionts, and the Origin of Mitochondria." *Bioscience Reports* 21 (1): 1–17.

Evans, N. 1996. *Extraterrestrial Life*. Edina: Burgess International Group.

Farmer, J. D. 2013. "Role of Geobiology in the Astrobiological Exploration of the Solar System." In *The Web of Geological Sciences: Advances, Impacts, and Interactions: Geological Society of America Special Paper 500*, edited by M. E. Bickford, Boulder, CO, 567–89.

Feng, D. F., G. Cho, and R. F. Doolittle. 1997. "Determining Divergence Times with a Protein Clock: Update and Reevaluation." *Proceedings of the National Academy of Sciences of the United States of America* 94 (24): 13028–33.

Fischetti, M., and J. Christiansen. 2021. "A New You in 80 Days." *Scientific American* 324: 4, 76.

Green, J. 2011. "Academic Aspects of Lunar Water Resources and Their Relevance to Lunar Protolife." *International Journal of Molecular Sciences* 12 (9): 6051–76.

Gura, C., and S. O. Rogers. 2020. "Metatranscriptomic and Metagenomic Analysis of Biological Diversity in Subglacial Lake Vostok (Antarctica)." *Biology* 9 (3): 55.

Kapitsa, A. P., J. K. Ridley, G. de, Q. Robin, M. J. Siegert, and I. A. Zotikov. 1996. "A Large Deep Freshwater Lake beneath the Ice of Central East Antarctica." *Nature* 381 (6584): 684–6.

Koonin, E. V., A. E. Gorbalenya, and K. M. Chumakov. 1989. "Tentative Identification of RNA-Dependent RNA Polymerases of dsRNA Viruses and Their Relationship to Positive Strand RNA Viral Polymerases." *FEBS Letters* 252 (1–2): 42–6.

Kumar, K., R. A. Mella-Herrera, and J. W. Golden. 2010. "Cyanobacterial Heterocysts." *Cold Spring Harbor Perspectives in Biology* 2 (4): 315.

Litschel, T., B. Ramm, R. Maas, M. Heymann, and P. Schwille. 2018. "Beating Vesicles: Encapsulated Protein Oscillations Cause Dynamic Membrane Deformations." *Angewandte Chemie International Edition* 57 (50): 16286–90. GDCh. Accessed 30 September 2018. https://doi.org/10.1002/anie.201808750.

Margulis, L., and D. Sagan. 1986. *Origins of Sex. Three Billion Years of Genetic Recombination*, 69–71, 87. New Haven: Yale University Press.

Miller, Stanley L. 1953. "Production of Amino Acids Under Possible Primitive Earth Conditions." *Science* 117 (3046): 528–9.

Nasir, A., and G., Caetano-Anollés. 2015. "A phylogenomic data-driven exploration of viral origins and evolution." *Sci Adv.* 1 (8): e1500527.

Oró, J., and A. P. Kimball. 1961. "Synthesis of Purines under Possible Primitive Earth Conditions. I. Adenine from Hydrogen Cyanide." *Archives of Biochemistry & Biophysics* 94: 217–27.

Oswald, G. K. A., and G. de, Q. Robin. 1973. "Lakes beneath the Antarctic Ice Sheet." *Nature* 245 (5423): 251–4.

Pelletier, J., L. Sun, K. Wise, N. Assad-Garcia, B. Karas, T. Deerinick, T. Ellison, M. Ellisman, A. Mershin, N. Gershenfeld, R.-Y. Chuang, J. Glass, and E. Strychalski. 2021. "Genetic Requirements for Cell Division in a Genomically Minimal Cell." *Cell* 184 (9): 2430–40. https:www.cell.com/action/showPdf?pii=S0092-8674%2821%2900293-2.

Powell, Kendall. 2018. "How Biologists Are Creating Life-Like Cells from Scratch." *Nature*. Accessed November 7, 2018. https://www.nature.com/articles/d41586-018-07289-x.

Schiermeier, Q. 2013. "Claims of Lake Vostok Fish Get Frosty Reponse." *Nature News*, July 9 2013. DOI:10.1038/nature.2013.13364.

Torres, Sosa, E. Martha, Juan P. Saucedo-Vázquez, and Peter M. H. Kroneck. 2015. "Chapter 1, Section 2 'The Rise of Dioxygen in the Atmosphere'." In *Sustaining Life on Planet Earth: Metalloenzymes Mastering Dioxygen and Other Chewy Gases*, edited by Peter M. H. Kroneck and Martha E. Sosa Torres, 1–12. Metal Ions in Life Sciences. 15. Springer.

Woes, C., Kandler, O., and Wheelis, M., 1990. "Towards a natural system of organisms Proposal for the domains archaea, bacteria, and eucarya." *Proceedings of the National Academy of Sciences, USA*, 87, 4576–4579.

CHAPTER 7

Evolution of Life and Intelligence

PROLOGUE

If you were part of an alien survey team visiting Earth two billion years ago, you would find no macroscopic lifeforms – no trees, no grass, no animals, no fish, not even an algae bloom. All life would be microscopic, composed of single-celled prokaryotic and early eukaryotic lifeforms. This may well be what human survey teams will find on nearby worlds. Even so, they would not be disappointed, for that would mean life has arisen independently in another world and, through its study, could lead to breakthroughs in many areas of research, including biochemistry, genetics, medicine, food production, and bioenergy. Perhaps more significantly, such a discovery would indicate that the occurrence of life on Earth is not unique, and we are likely to find it in various evolutionary states throughout the cosmos.

7.1 INTRODUCTION

Since the first eukaryotic cells appeared in the primordial seas ~2 billion years ago, there has been an explosion in the diversity and level of complexity of lifeforms on Earth. This push toward biodiversity was driven by variations in the natural environment, increased reproductive rates resulting from the use of oxygen as a metabolic catalyst, and the underlying principles of biological evolution. In his book *On the Origin of the Species* (1859) Charles Darwin postulated that when a species undergoes a potentially fatal environmental stress, only those individuals (if any) which, by chance, are genetically predisposed to handle the situation will survive. Those surviving individuals will then be able to utilize the resources at hand to reproduce and pass their life-saving genetic traits on to their offspring, thereby increasing the immunity of subsequent generations to the threat. The underlying principles of evolution by natural selection are (1) that genetic differences within a species can lead to individuals having varying traits, (2) traits are inheritable, and (3) there are environmental stresses that lead to a struggle for existence. Genetic differences within

DOI: 10.1201/9781315210643-7

a species lead to variations in traits, while natural selection works to sort the variations into those that are lifesaving and those that are not. Amazingly, Darwin's insights into the importance and influence of genetic variations on life predated the discovery of DNA. DNA provides the physical mechanism through which genetic variations can occur and be expressed. Organisms with the same or similar DNA sequence are said to be of the same or similar genotype.

7.2 GENETIC VARIATIONS

Genetic variations between members of a species are caused by variations in the nucleotide sequences between individuals. Depending on the extent of the variations they may be manifested in outwardly observable characteristics, i.e., phenotype. Variations over many genes can lead, for example, to variations in body type or intelligence, while differences in the genetic coding of just a few genes can lead to changes in such things as blood type or eye color. The variant form of a given gene is referred to as an allele. There may be two alleles for a given gene (i.e., a small section of DNA). The interaction between them determines which allele is dominant and which is recessive. Some genes express themselves no matter what, others only under certain environmental circumstances. For example, Himalayan rabbits have a gene for producing black fur when they are reared at high altitudes where temperatures can drop below 35°C; otherwise their fur remains white. Black fur is a better absorber of ultraviolet light, which helps to keep the rabbit warm. The first parts of the rabbit that reaches this temperature are its ears and feet, which turn black in the winter (Sturtevant 1913; Lobo 2008).

7.2.1 Random Mutations

Over the history of the Earth the principal source of genetic variation has been random mutation within nucleotide sequences (see Figure 7.1). These can be caused by a number of factors; these include DNA-copying errors, cosmic rays, radioactivity, ultraviolet light, and chemical toxins. The impact of a mutation depends on where it happens in the DNA. DNA contains both coding and noncoding sequences. Coding sequences are those that are responsible for specifying and directing the creation of proteins. Perhaps only 8–15% of DNA serves this function (Ponting and Hadison 2011; Rands et al. 2104). If mutations occur within the protein-coding part of DNA, it is estimated that ~70% of the time they will lead to damaging results (Sawyer et al. 2007). Noncoding DNA, while not responsible for specifying proteins, stores knowledge on how to make various forms of RNA used by the cell (tRNA and rRNA), can regulate DNA transcription, and impact gene function. In total, at least 80% of the human genome is believed to be biologically active (ENCODE 2012). This would suggest that <20% of your DNA is inactive or "junk DNA" (Pennisi 2012).

As a segment of DNA is being read to instruct the cell on what to do (see Figure 6.19), it is important that the "junk DNA" be avoided. For this reason, there are special "stop" base sequences (a codon) that prevent the DNA from being read past the good base pairs into either the junk base pairs or base pairs associated with another protein. Cancer occurs

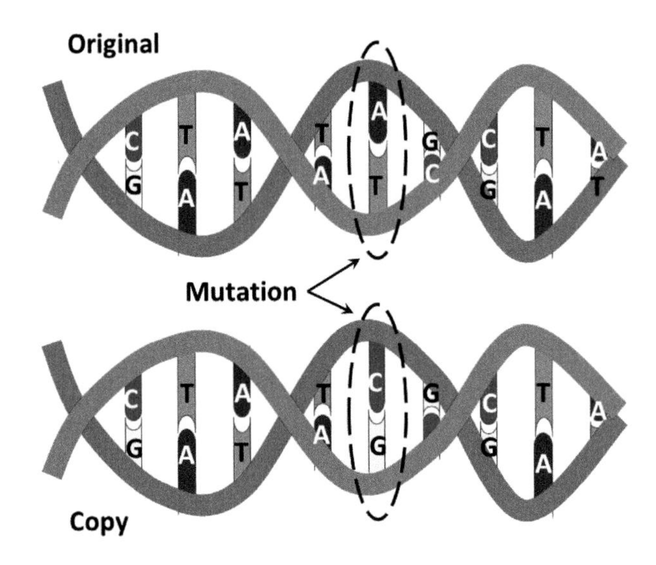

FIGURE 7.1 Mutation in DNA. A strand of DNA is said to be mutated when its copy has one or more changes in its nucleotide sequence compared to the original. Mutations are largely random occurrences that can be caused by copying errors, toxic chemicals, UV light, radioactivity, or cosmic rays. Once a mutation is present, it can be handed down to subsequent generations. The mutation of even one nucleotide can have a significant impact. For example, sickle cell anemia is caused by the switch of an "A" to a "T" base in a codon sequence (GAG to GTG) within the Beta-globin gene.

when cells divide and the "stop" instructions are ignored. In this case excess proteins are produced, resulting in a tumor.

The fact that most mutations have occurred randomly is a powerful thing. This means their occurrence was governed by the laws of probability and not by the current need of the organism. Therefore, such mutations, while not useful at the moment, may prove extremely valuable to an individual due to some unexpected future event. Let us consider the case of bubonic plague. In 1347 AD, the bubonic plague (i.e., the Black Death, Figure 7.2) began to sweep across Europe, sometimes decimating the population of an entire region. Ultimately, plague killed ~30% of the world's population, the impact of which persists in some parts of Europe even to this day. Plague is a bacterial infection spread by the bite of a flea. Given the poor sanitary conditions present during the Middle Ages, fleas were common place in human society. Before 1347 AD the plague was largely unknown in Europe. Any natural immunity to it an individual may have harbored had been of no consequence. Of course, once plague arrived, this was no longer the case. Fortunately, about half the population had DNA with beneficial genetic mutations that endowed them with the ability to survive infection.

7.2.2 The Importance of Sex

Since mutations are often more harmful than not, the process of natural selection has worked to decrease the number of random mutations associated with replicating DNA from one generation to the next. In simple viruses, the base pair error rate is 1 in a 1,000.

FIGURE 7.2 The Black Death. Conveyed by fleas, the bubonic plague, "Black Death", swept through Western Europe beginning in 1347. Before it was over ~1/3 of the population was killed. Due to previously unimportant, random genetic mutations, some individuals were better able to resist its effects than others. Image Credit: Shutterstock.

Prokaryotic and eukaryotic cells have developed special proofreading enzymes that check each base pair to ensure the correct purine base is connected to the correct pyrimidine base. If an error is found, it is corrected. Mutations from copying errors occur only when this mechanism fails. With this proofreading system the base pair error rate is improved by a factor of 1,000 in bacteria and an additional factor of 1,000 in eukaryotic cells, resulting in an error rate of only ~1 in a billion base pairs.

This low error rate is both good and bad. It is good because it helps ensure a successful organism can reliably reproduce itself. A low mutation rate can be bad because it dictates that evolution relying principally on copying errors for genetic diversity is an extremely slow process, potentially leaving organisms vulnerable to extinction if there are no individuals within their population genetically equipped to handle the next unforeseen crisis.

For much of Earth's history evolution was at a near standstill, governed by the slow rate of random mutations. The offspring of organisms were, to within copying errors, genetically identical to their parents – clones. Reproduction of this type with a single genetic parent is termed "asexual". But then, about 1.2 billion years ago (Butterfield 2000) the culmination of random mutations, together with environmental pressures, led to the evolution of the first sexual, eukaryotic organisms. In sexual reproduction there are two parents, each contributing half the chromosomes to the offspring via specialized reproductive cells called gametes (egg from the female and sperm from the male) that combine in a single-celled zygote through a cross-over process. Cell division then faithfully copies the new DNA within a multicellular embryo.

The number of chromosome pairs within a cell varies from organism to organism. Humans have 23 pairs, for a total of 46 chromosomes (23 chromosomes from each parent). The zygote represents the first cell of the offspring. This new DNA base pair sequence is then faithfully copied through cellular division to all cells within the growing offspring. Exactly how the pairs combine will vary from offspring to offspring, even when the parents

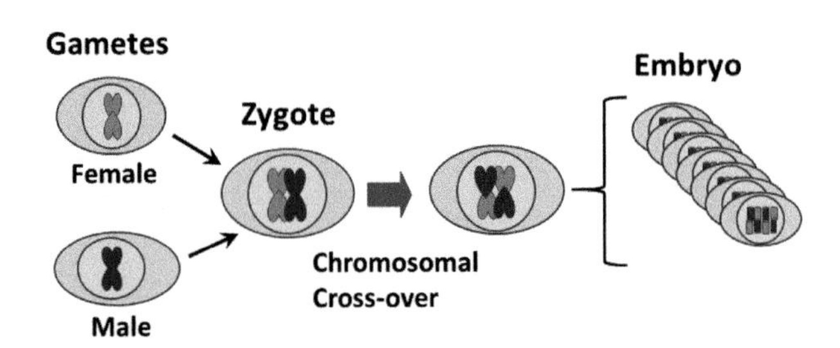

FIGURE 7.3 Sexual reproduction. Sexual reproduction is a process that enables the DNA of offspring to contain the coding sequences of two parents. Each parent contributes a gamete containing half the parent's chromosomes to the process. When the gametes combine a zygote cell results. The zygote is the first cell of the offspring. Within the zygote the chromosome pairs are blended in a process called "cross-over". Which coding sequences for a given parental trait are expressed in the resulting DNA is up to chance, making each offspring of the same two parents a unique genotype, except when the offspring arise from the same zygote (e.g., in the case of identical twins). Cell division then conveys this new, unique genetic code to the growing, multicellular embryo.

are the same. An exception to this is the instance of multiple births where the offspring originate from the division of the same zygote (e.g., identical twins).

Now, through sexual reproduction, life on Earth could enjoy a great wealth of genetic diversity, giving it a better chance of dealing with whatever hardship lay ahead. It also allowed life to expand into what were once inhospitable environments. Another advantage of sexual reproduction is that it prevents the accumulation of genetic mutations within a single organism. However, sexual reproduction does come at a cost. While all asexual organisms can reproduce, it takes two sexual organisms (male and female) to reproduce, giving asexual populations the ability to reproduce exponentially faster than a sexual population. Also, only half the genetic information from any one sexual individual is passed on to its offspring (Figure 7.3).

The vast majority of today's animal species undergo sexual reproduction, while plants are mostly asexual, but many can also reproduce sexually. The number of animal species (~8 million) greatly exceeds the number of plant species (~300,000) and other (~700,000; fungi, protozoa, and chromista) eukaryotic species. However, if one measures genetic success by weight, asexual prokaryotes are the big winners – having an estimated total biomass >100 times that of eukaryotes (Whitman, Coleman, and Wiebe 1998). Even after all this time, we are still a planet dominated by primitive, single-celled lifeforms.

7.3 STRUGGLE TO SURVIVE

The day-to-day struggle to survive provides a constant pressure that drives the evolution of life forward. Creatures that can best cope with the environment in which they find themselves are able to utilize natural resources and procreate, carrying the information stored in their nucleic acids forward in time. Compared to many other worlds that have recently been discovered, the Earth is a "life-friendly planet" (Figure 7.4). It is in a stable orbit about a serene, middle-aged, G4 yellow dwarf star that bathes the Earth

FIGURE 7.4 Mother Earth. The Earth is a "life-friendly" world. It is in a stable orbit at just the right distance from a long-lived, stable Sun. The Earth's mass is sufficient to retain an atmosphere and to support, low-level, plate tectonic activity. An unusually massive Moon helps keep the Earth from wobbling on its rotation axis and drives oceanic tides. Credit: NASA.

with sufficient energy to keep the surface temperature largely between the freezing and boiling points of water. The Earth rotates on its axis and is kept from precessing wildly due to the stabilizing effects of a large orbiting Moon. The Earth's rotation spreads the warmth of the Sun evenly over its surface, softening the temperature swings between day and night. The Earth also generates its own internal heat that continually drives the conveyor belts of plate tectonic and volcanic activity, which serve as relief valves against the possibility of a dangerous build-up of internal pressure. The Earth's atmosphere is believed to have originated largely from the outgassing of volcanoes, with most of its CO_2 being washed out by rains, thereby avoiding a runaway greenhouse effect like is occurring on Venus. Enough atmospheric CO_2 was left behind to provide a blanket to keep the Earth's surface warm. Under these steady-state conditions, the slow rate of mutations occurring from generation to generation in primitive prokaryotic lifeforms was sufficient to gradually push life forward. A slow gradual evolution of life is just what classical Darwinian evolution expounded. However, there is little evidence in the fossil record to back up the concept of evolutionary "gradualism". Instead, what is found are long periods of slow, gradual changes consistent with random mutations in a benign environment. But then, out of nowhere, there are sudden, rapid changes in speciation followed by yet another long stretch of gradual evolution. Such an evolutionary trend – long periods of gradual evolution (called stasis) followed by abrupt shifts – is referred to as "punctuated equilibrium" (Eldredge and Gould 1985; Figure 7.5). The shifts are caused by sudden, dramatic changes in environmental conditions. The greater the change, the less likely an organism would be able to survive. Indeed, the fossil record shows that if the changes are sufficiently severe, they will result in massive death. In such instances, they are referred to as extinction events.

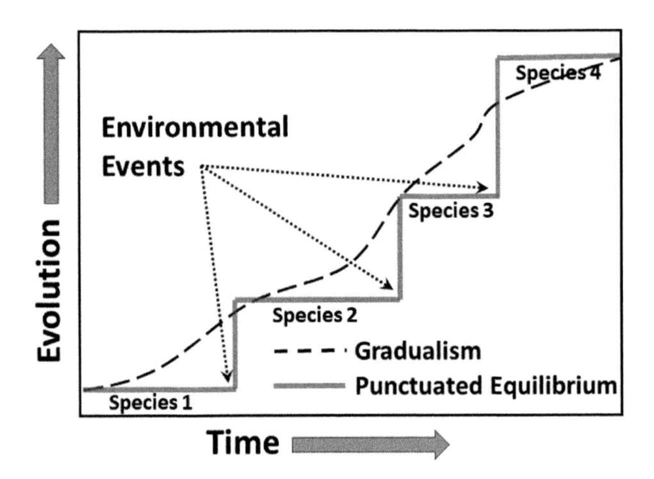

FIGURE 7.5 Theories of evolution. Darwinian gradualism postulates evolution occurs gradually over time due to random mutations and continuous environmental pressure. Punctuated equilibrium theorizes that evolution occurs in fits and starts due to occasional rapid changes in environmental conditions to which species possessing favorable random, inheritable mutations are able to survive and evolve due to the availability of expanded habitat. Evolution occurs so rapidly that transitional species are seldom found.

7.4 MASS EXTINCTION EVENTS

As described in Chapter 6, the first extinction event is widely believed to have been the Oxygen Crisis, occurring ~2.45 Bya (billion years ago). This is the time when oxygen levels in the Earth's atmosphere reached significant levels for the first time. The enhanced levels resulted from the build-up of photosynthetic cyanobacteria in the oceans. Before this time all life on Earth was anaerobic, optimized for life in an oxygen-free environment. For these creatures, oxygen was a toxic gas.

There have been at least 25 mass extinction events since the Oxygen Crisis. Five of these are considered to have been major and are described in Table 7.1 (Sepkoski and Raup 1982; Sepkoski 1984; Alroy 2008). Mass extinctions reset the evolutionary clock on Earth, clearing out old species and allowing space for new or less successful species to evolve. This was the case 66 Mya after the Cretaceous–Paleogene event (previously referred to as the K-T event; Figure 7.6) when three-quarters of all plant and animal life on Earth were wiped out due to the impact of a ~10 mile diameter comet or asteroid off the coast of what is today the Yucatan Peninsula. The event is evidenced by a thin (~1 cm thick) boundary layer of iridium found all around the world. Iridium is extremely rare on Earth, but abundant in asteroids and comets (Alvarez and Alvarez 1980). The energy released during this event was roughly equivalent to 1,000 times the power of all nuclear weapons on Earth going off at once. The event devastated the dinosaurs and affected ~75% of all species. Fortunately for us, one of the survivors was a previously unremarkable mammalian species which flourished and rapidly evolved to fill the ecological niches vacated by its deceased competitors.

It is interesting to note that all but one of the major extinction events since the Oxygen Crisis are astronomical in origin; three due to known or suspected meteor or comet strikes and the fourth suspected of being due to the radiation from a Gamma Ray Burster (GRB)

TABLE 7.1 Mass Extinction Events

Event	Time	Suspected Cause	Extinction	Reference
Ordovician-Silurian	450–440 Mya	Gamma Ray Burster	70%	Melott et. al. 2004
Late Devonian	375–360 Mya	Magmatism/Volcanism	70%	Ricci et al. 2013
Permian-Triassic	252 Mya	Meteor/Comet strike	90%	Von Frese et al. 2009
Triassic-Jurassic	201.3 Ma	Meteor/Comet strike	70 - 75%	Blackburn et al. 2013
Cretaceous-Paleogene	66 Mya	Meteor/Comet strike	75%	Randall 2015

FIGURE 7.6 C-P event. Approximately 66 million years ago a massive, ~13 km, wide comet or asteroid impacted the Earth off the coast of what is today the Yucatan Peninsula. The energy release was equivalent to ~1,000 all out nuclear wars, which decimated much of life on Earth, including the dinosaurs. With the extinction of the dinosaurs, minor species, such as mammals, were able to expand their domains. Image: NCAR.

going off within about 10,000 light years of the Sun. GRBs are believed to be due to the collisions of neutron stars, which is expected to be a rare occurrence. However, this is not the case with meteor impacts. Figure 7.7 is a plot of the expected frequency of meteor and/or comet impacts with the Earth as a function of time. An impact of an asteroid/comet the size of the one that took out the dinosaurs is expected to occur every ~50 million years or so. We are statistically overdue.

7.5 FOSSIL RECORD

The history of life on Earth has been deduced from the fossil record. A fossil is any trace or remnant of a once-living organism left behind in the sediments or rocks of the Earth. Fossils can range in age from ~10,000 to 4.1 billion years old (Bell et al. 2015) and be as small as a microbe or as large as a dinosaur. Examples of fossils include impressions left in dried mud, solidified insects trapped in amber (see Figure 7.8), and petrified or permineralized wood, bones, and shells. Premineralization (Mani 1996) occurs when minerals dissolved in water permeate an organism's pores and tissue. Once the water evaporates and the minerals solidify, an accurate representation of the organism is left behind. Petrification occurs when all organic material is replaced by minerals. In the nineteenth century, the

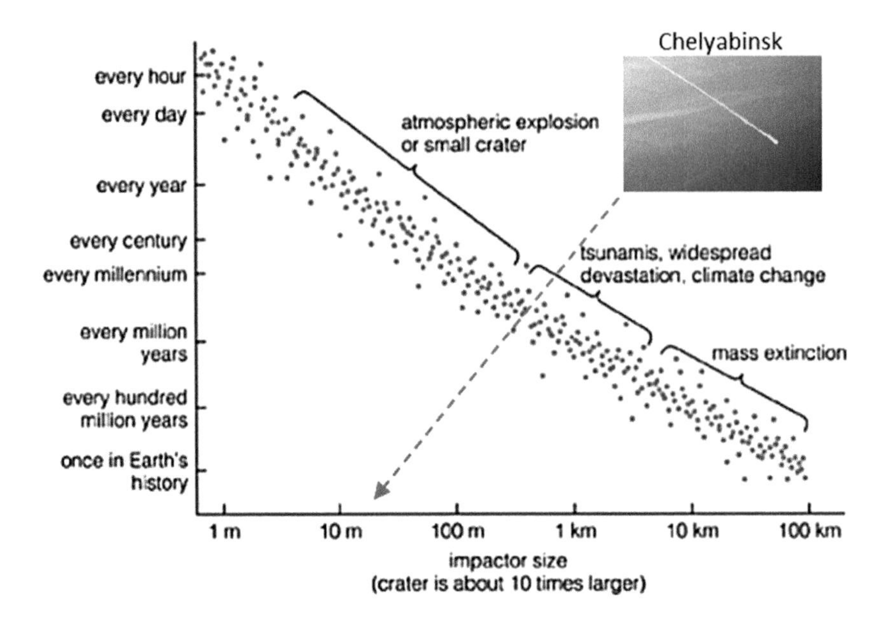

FIGURE 7.7 Frequency of occurrence versus size of impactor. Over its 4.6 billion years of existence, the Earth has frequently endured collisions with foreign bodies, everything from small meteoroids responsible for "shooting stars" to a Mars size body that nearly destroyed the Earth and was responsible for the formation of the Moon. Approximately 66 million years ago a ~10 km diameter asteroid struck the Earth with the equivalent firepower of a ~10 Megaton nuclear bomb, taking out the dinosaurs and a host of other species. The Tunguska event refers to the explosion of a ~100 m object ~24,000 ft over eastern Siberia on June 30, 1908. The explosion had the energy release of a 10-15 Megaton nuclear bomb and flattened trees over 2,000 km^2 region. The Chelyabinsk meteor was ~20 m diameter object that exploded in an airburst ~97,000 ft over Russia on February 15, 2013, with the energy of a 400–500 kiloton nuclear bomb. Credit: NASA/JPL: Inset Shutterstock.

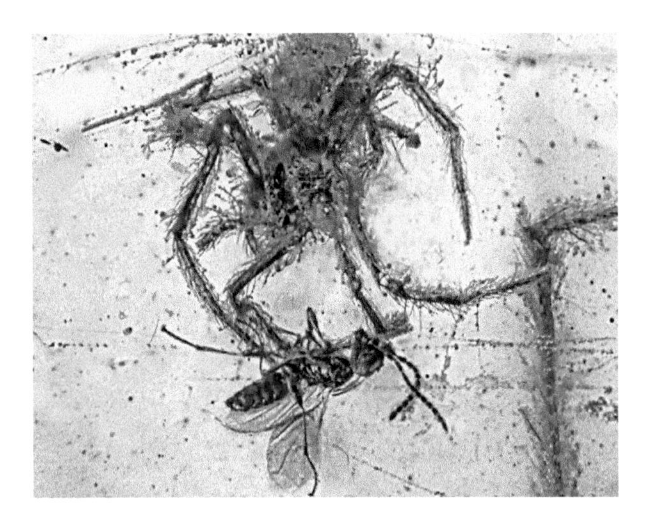

FIGURE 7.8 Eternal struggle. The battle between a wasp and spider frozen in time within fossilized tree sap (amber). The fossil dates from the Cretaceous period between 97 and 100 million years ago. Found in the Hukawng Valley of Nyammar. Image Credit: George Poinar, Oregon State University, with permission.

age of fossils was deduced from knowledge of where they occurred in the sediment and the rate of material deposition. This permitted fossils to be dated back some millions of years. A much more accurate approach that can be used all the way back to the formation of the Earth is radiometric dating.

7.5.1 Radiometric Dating

Radiometric dating makes use of the fact that some elements are inherently unstable and can spontaneously eject alpha particles (2 protons + 2 neutrons), beta particles (1 electron or positron + neutrino), an electron, or a gamma ray. When an alpha particle is ejected, the number of protons in the emitting substance changes, thereby changing its elemental identity. For example, uranium 235 (^{235}U; used to make fission bombs) decays into lead 207 (^{207}Pb). The ^{207}Pb is said to be a daughter radioactive isotope of the parent isotope ^{235}U and is itself unstable and subject to further decay. The exact moment at which a given atom of ^{235}U will decay into ^{207}Pb is unknown, but the rate at which a collection of radioactive nuclei will decay into the next radioactive isotope is known to follow an exponential function governed by the decay half-life (Soddy 1913). The half-life is the time (typically expressed in years), it will take for one half the number of radioactive isotope nuclei in a sample to decay into its daughter isotope. Therefore, the ratio of the parent to daughter isotopes in a given sample can be used to estimate the sample's age. In the case of the $^{235}U \longrightarrow {}^{207}Pb$ decay, the half-life is 710 million years. In 1907 Bertram Boltwood first noted that the lead-to-uranium ratio was greater in older rocks (Boltwood 1907). At the suggestion of Ernest Rutherford (1905), Boltwood then used the ratio to estimate the age of a uranium rock sample (see Figure 7.9). This was the first known use of radiometric dating. With the invention and further refinement of the mass spectrometer (Dempster 1918; Dickin

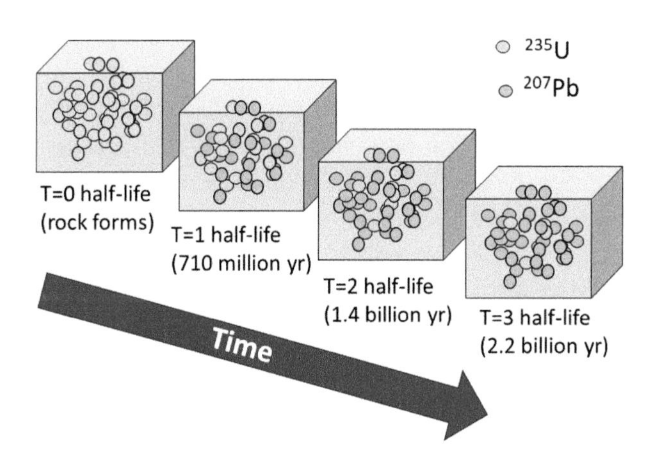

FIGURE 7.9 Radiometric dating. Over its half-life of 710 million years half the uranium-235 (^{235}U) atoms in a sample will decay into lead-207 (^{207}Pb) atoms. In two half-lives (1.4 billion years) half of the remaining (^{235}U) atoms will also decay into ^{207}Pb. This decay process will continue in an exponential manner until no more ^{235}U atoms are left in the sample or to the time of the Earth's destruction, whichever comes first.

TABLE 7.2 Half-lives of Radioactive Isotopes

Isotope		Half-life	Useful Range
Parent	Daughter	(years)	(years)
Carbon 14	Nitrogen 14	5,730	100–60,000
Potassium 40	Argon 40	1.3 billion	100,000–4.5 billion
Uranium 235	Lead 207	710 million	10 million–4.6 billion
Uranium 238	Lead 206	4.5 billion	
Rubidium 87	Strontium 87	48.8 billion	10 million–4.6 billion

2008), techniques for determining the ratio of parent to daughter isotopes have continued to improve, allowing ever more accurate determinations of sample ages to be made.

A listing of radioactive isotopes commonly used in radiometric dating is provided in Table 7.2. The long half-lives of radioactive isotopes of potassium, uranium, and rubidium make them well suited for dating geologic samples. In contrast, the relatively short half-life of carbon-14 (5,730 years) makes it ideal for use in dating archeological finds and the remains of plants and animals. Carbon-14 dating (i.e., radiocarbon dating) was pioneered in the 1940s by William Libby and his colleagues (Arnold et al. 1949). On Earth most carbon-14 is generated in the Earth's upper atmosphere (30,000 to 49,000ft) through the interaction of cosmic rays with atmospheric nitrogen-14. The carbon-14 then combines with atmospheric oxygen to make $^{14}CO_2$, which is incorporated into plants through photosynthesis. The plants are then eaten by herbivores which can subsequently be eaten by carnivores. In this way carbon-14 is consumed by and exists in similar concentrations within many lifeforms. Once a lifeform dies the carbon-14 stopwatch begins, as the carbon-14 atoms present at the time of death begin their decay back to nitrogen-14 with little or no chance of replenishment. From this point on, the time since death can be estimated from the ratio of carbon-14 to nitrogen-14 until there is too little carbon-14 left to accurately measure. The time limit for the use of carbon-14 dating is ~60,000 years. In the above discussion, it was assumed the amount of carbon-14 in the Earth's atmosphere has remained constant over time. However, it is known the amount of carbon-14 has varied significantly, requiring a correction be made to the derived ages using data from other sources, for example from the isotopic analysis of ancient air bubbles trapped in polar ice cores (Wilson and Donahue 1992; Aitken 1990).

7.5.2 Timeline for Life on Earth

Figure 7.10 shows a timeline for the evolution of life on Earth. Radiometric dating of rocks from Eastern Australia suggests that simple, single-celled, archaea lifeforms may have arisen on Earth within a few hundred million years of its formation – soon after the first pools of water collected on the newly formed surface (Bell et al. 2015). In these early times, there was no free oxygen in the Earth's atmosphere, and all life was anaerobic. After a billion years, a series of chance mutations led to the ability of some bacteria (e.g., cyanobacteria) to use water, atmospheric carbon dioxide, and light to create carbohydrate molecules and molecular oxygen (O_2). The availability of energy-rich carbohydrate molecules helped to boost cellular metabolism and reproduction, driving evolution forward. Over the next

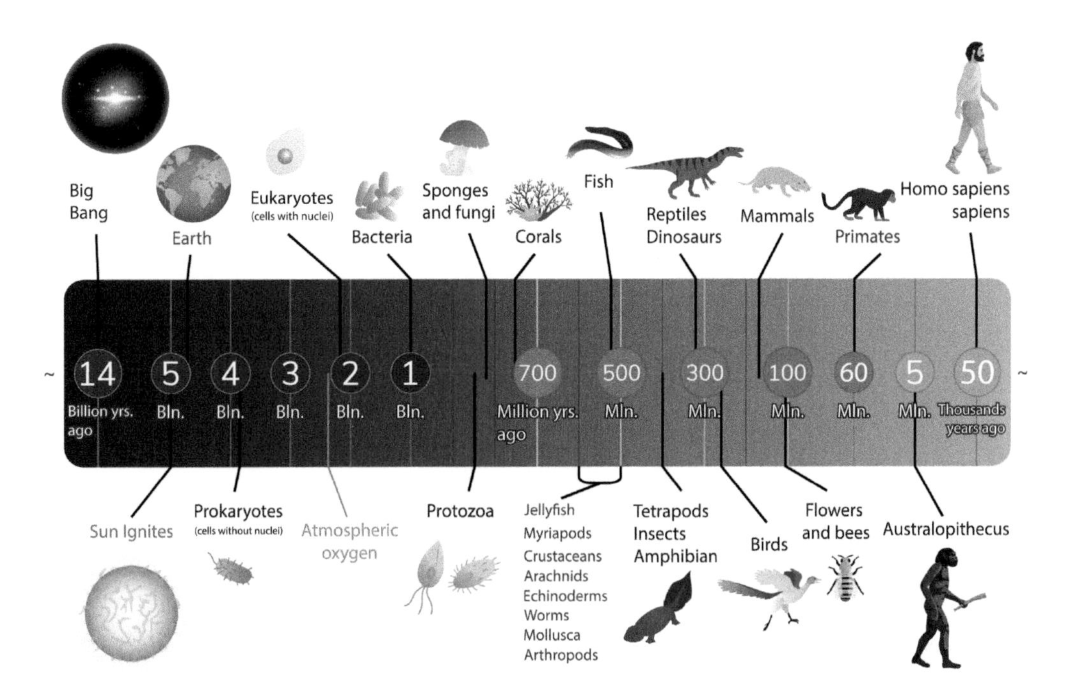

FIGURE 7.10 Timeline for life on Earth. The fossil record indicates life began within the first billion years after the Earth's formation ~4.6 billion years ago. Outgassing from volcanoes provided an atmosphere of carbon dioxide, nitrogen, and water vapor, but no oxygen. The first lifeforms were simple prokaryotes, with no nucleus. They likely used strands of RNA to control their metabolism and convey genetic information from one generation to the next. Over the next billion years random, inheritable mutations led the evolution of cyanobacteria capable of photosynthesis. The oxygen that was released by the cyanobacteria was toxic to most lifeforms. However, some adapted and used it to "super-charge" their metabolism leading to an acceleration in evolution and the realization of more complex eukaryotic and multicellular organisms. Rising atmospheric oxygen levels together with increasing genetic complexity and other environmental/ecological factors led to the Cambrian Explosion where the plant and animal lifeforms that are familiar to us got their start. For much of its history the most advanced lifeforms on Earth were unicellular. Only in comparatively recent times have complex plant and animal lifeforms come to exist. Credit: Shutterstock.

billion years, the biogenically produced O_2 was chemically captured by iron or organic matter in the ocean. Once these oxygen sinks were full, the oxygen released by the cyanobacteria was able to collect in the atmosphere until it reached toxic levels to the exposed anaerobic lifeforms (Figure 7.11). This was one of the most devastating mass extinction events in Earth's history. As described in Section 6.4, some lifeforms were able to adapt to the presence of O_2 and use it within chloroplasts and mitochondria to further their evolution; first to the complexity of eukaryotic cells and then to multicellular lifeforms. As can be seen in Figure 7.11, the percentage of O_2 in the Earth's atmosphere just past 10% at the start of the Cambrian Explosion, about 542 million years ago. The Cambrian Explosion is considered by many to have been the biological equivalent of the "Big Bang" for life on Earth. Within 70–80 million years of the onset of this event, there was a rapid acceleration in biological diversity as witnessed by the fossil record. Early versions of all

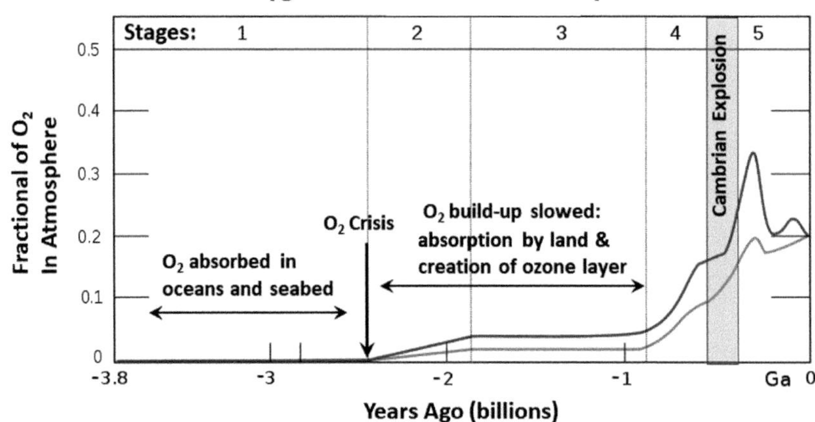

FIGURE 7.11 Accumulation of atmospheric O_2. The red and green curves indicate approximate upper and lower limits to the molecular oxygen (O_2) level. All O_2 within the Earth's atmosphere is produced biologically through photosynthesis. The production of O_2 began ~3.5 billion years ago with the evolution of cyanobacteria within the Earth's oceans. For nearly a billion years all the oxygen that was produced was chemically captured by dissolved iron and organic matter in the ocean, which acted as an oxygen "sink". Once the oxygen sinks were chemically saturated, O_2 began to freely percolate through the oceans and accumulate in the atmosphere. To most of the anaerobic lifeforms then in existence O_2 was a toxic gas. This led to the O_2 Crisis, where most of the lifeforms died off, leaving only the ones that could adapt to its presence. The build-up of O_2 in the atmosphere was slowed down by its chemical absorption by land masses and the build-up of the ozone layer. Once this second oxygen sink was filled, atmospheric oxygen levels continued to increase. Since the start of the Cambrian Explosion (see Figure 7.11 and text) O_2 levels have varied between 15 and 35%, compared to 21% today (Berner 1999). The peak occurred to the Carboniferous period and may have promoted gigantism in plants and amphibians that existed at that time. Higher atmospheric pressures and oxygen levels may also have increased the number of species capable of flight, as well as the occurrence of wildfires (Dudley 1998). Adapted from https://commons.wikimedia.org/wiki/User:Loudubewe

major animal body types that exist today developed within this narrow time frame (Figure 7.12). The body types correspond to Phyla in the Linnaeus classification system (including ours, Chordata). The establishment of these biological pathways set the course for the further evolution of life on Earth. Exactly what triggered the Cambrian Explosion is still the subject of debate. The atmospheric oxygen level could be one of several contributing factors. Other potential factors include the melting of glaciers, reaching a critical threshold in genetic complexity, and the emergence of predators (i.e., evolve or die).

7.6 THE EVOLUTION OF INTELLIGENCE

7.6.1 Genetic Intelligence

Evolution has resulted in an increase in the variety and complexity of life on Earth. Complexity carries with it a connotation of fragility. So, why would evolution (not always, but sometimes) select for it? The push toward greater complexity is driven by environmental

FIGURE 7.12 Fossil starfish. The starfish of the Phylum Echinodermata represents one of the body types that evolved during the Cambrian Explosion. The specimen shown was excavated in the Sahara Desert and is from the Ordovician Period, 488–433 million years ago. Photo by author.

pressures that over time provide an organism with additional attributes that improve its survivability, whether that be the ability to identify and acquire food sources or to adapt to changes in its environment. For most of the Earth's history these attributes could only be passed from one generation to the next through the information contained in the base sequences within an organism's DNA.

The basic unit of information in a computer is a binary digit (or "bit", for short), which can be either a "0" (no) or "1" (yes). As we learned in Chapter 6 (Figure 6.8), three base pairs (i.e., a codon) are sufficient to designate one of the 20 amino acids used by lifeforms on Earth. Each base pair contains two bits of information, so a codon contains six bits of information. Human DNA contains about 3 billion bases, corresponding to $\sim 6 \times 10^9$ bits (i.e., 6 gigabits) of information. This is the amount of information contained in 4,000 books, each 500 pages long (Evans 1996). This multivolume set contains all the information needed to construct a human being. This "Encyclopedia Human" is so important to our moment-to-moment existence that there is a copy of it contained within the nucleus of every cell in our bodies. This is also true for every lifeform on Earth and consistent with the idea that all life evolved from a common, single-celled ancestor. You are composed of ~30 *trillion* cells, which means you are personally carrying around about as many pages of genetic information as there are grains of sand in all the beaches and deserts of the Earth.

Our cells replicate about once every 24 hours. Each time a cell replicates it passes on a copy of its DNA. How much information can be stored in a strand of DNA is set by the replication error rate for a given organism. To ensure proper cell operation, the DNA should be faithfully copied to approximately the 99.999% level. In the case of a virus, the simplest and earliest form of life on Earth, the error rate is about 1 in every 1,000 base pairs. Therefore, the replication error rate limits the length of a virus DNA and, therefore, its information storage capability to ~10,000 bits. The survival pressure toward greater

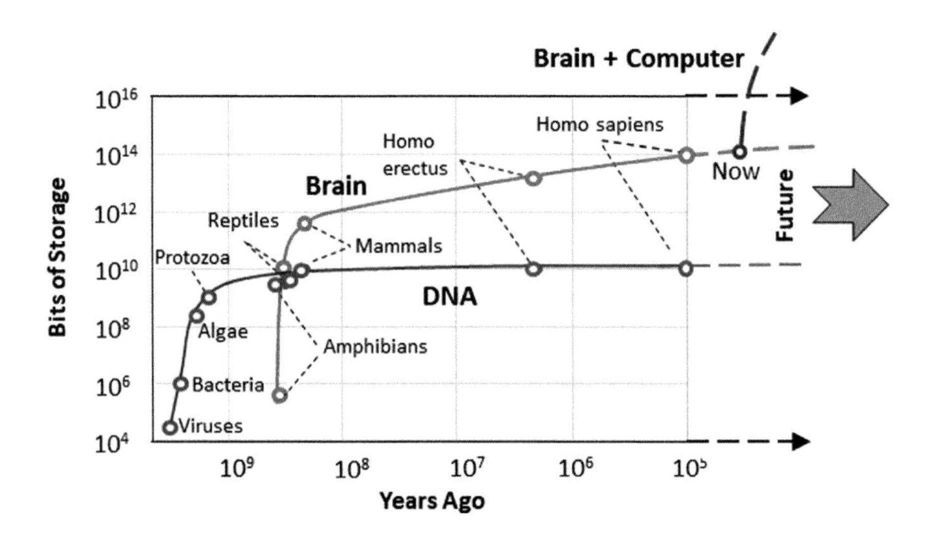

FIGURE 7.13 Evolution of information storage capacity. Initially, the only means life had for storing information was in the base sequences of nucleic acids, e.g., DNA. The amount of data that can be stored in DNA is set by the replication error rate, which approached peak efficiency around a billion years ago, leading to a data storage capacity of ~10 gigabits. Soon after evolutionary pressure led to the appearance of the first brains, which rapidly evolved and allowed both increase storage capacity and the ability for an organism to both acquire and utilize knowledge within its own lifetime. The storage capacity of brains is now leveling off, but is being augmented by electronic devices. Adapted from Sagan 1977 and Evans 1996.

complexity to achieve survivability soon (within 200 million years or so) led to the evolution of special error correcting proteins (i.e., enzymes) that reduced the error rate to just one in a million (or 10^{-6} in scientific notation), allowing DNA strands up to 10 million times longer to be faithfully reproduced. This length of DNA was sufficient to provide the information required for more advanced prokaryotic, bacteria-like organisms to exist and reproduce. Over the past few billion years the error-correcting mechanisms within the nuclei of cells have evolved to the point where the error rate has dropped to one in a billion base pairs, or 10^{-9}, pushing the potential storage capability within modern eukaryotic cells to 10^{10} bits (Evans 1996). This evolution toward greater information storage capacity within nucleic acids is illustrated in Figure 7.13. As can be seen in the figure, the DNA-storage capacity has leveled off over the past billion years, suggesting that the capability of DNA to self-correct has reached its practical limit. However, quoting a well-known line in the film *Jurassic Park*, "Life will find a way".

Information stored in DNA is "hardwired", like the boot-up program on your computer, and can only change due to mutations or sexual reproduction between generations. Often times these changes have little or no impact. However, sometimes they can provide a new attribute which provides an organism an enhanced chance at survival and reproduction over its contemporaries, in which case the beneficial attribute will be passed down to its offspring. This is a slow and painful way for organisms to advance up the evolutionary ladder.

7.6.2 Brain Intelligence

The turnover in DNA-storage capacity occurred at about the time of the Cambrian Explosion ~560 million years ago (see Figure 7.10), when there was an evolutionary "big bang" resulting in a wide variety of new lifeforms. The occurrence of the Cambrian Explosion suggests that by this time evolutionary pressure had pushed the information storage capacity of DNA past the genetic threshold needed to allow the proliferation of many species. One of these species, an amphibian, contained a rudimentary brain capable of storing information outside of DNA. Unlike the information stored in DNA, the information stored in the newly developed brain could be altered to enhance the chance of survival within an individual organism's lifetime, increasing its prospects of reproduction. This is a *huge* evolutionary advantage for the individual organism. A creature with a sufficiently large brain can more rapidly adapt to changes in the environment, e.g., avoiding predators and finding mates, not to mention food and shelter. The DNA in organisms without brains can, to a limited extent, express itself differently depending on changes in environmental conditions (e.g., flowers blooming in the spring), but can only adapt to unexpected changes in environment over generational time scales. However, if you rate evolutionary success by biomass, creatures without brains (e.g., plants, bacteria, and protists) are the big winners, with creatures harboring brains accounting for less than 0.4% of life on Earth. It takes far more planetary resources to support a population of organisms with brains than it takes to support an equivalent mass of organisms without brains. Although such organisms win in terms of biomass, they respond only the instinctual calls of their cellular DNA and pass through their lives with no awareness of their own existence.

The Oxford Dictionary defines intelligence as "the ability to acquire and apply knowledge and skills". Although a DNA molecule is not self-aware, it can, over multiple generations, acquire and apply the information it contains to improve the survivability of its host cell. So, perhaps it is fair to ascribe to DNA a degree of "molecular intelligence". However, the ability to acquire and apply knowledge within one's own lifetime requires an organism to be self-aware and for that you need a brain. As can be seen from Figure 7.13, the competitive edge in the survival game provided by having a brain increased the chances for those organisms to reproduce and, consequently, drove their rapid evolution. Indeed, it took less than 100 million years for the storage capacity of brains to catch up to and surpass that of the venerable DNA molecule.

Brains are composed of neuronal and glial cells. An average human brain contains ~100 billion neurons. Neurons are the smallest information-processing unit in a brain. They use electrical impulses to process and convey information and are analogous in function to transistors in a computer. Glial cells do not process information themselves, but serve a role analogous to a computer's chassis, providing a stable life support system (e.g., electrical/mechanical isolation between neurons, nutrients/power, thermal management), without which neuronal networks could not function. Neurons make up 60–80% of a brain's mass, with the remaining 20–40% of the mass in glial cells (Mota and Herculano-Housel 2014).

There are three types of neurons in the human body: (a) afferent neurons, (b) efferent neurons, and (c) interneurons. Afferent neurons in your skin, eyes, and ears provide sensory input (e.g., heat, light, and touch) to the brain. Efferent neurons run through the spinal cord and carry control signals from the brain to muscles, glands, and organs. Interneurons connect one neuron to another and are the most abundant type of neuron in the body. Efferent neurons can be up to a meter long, while interneurons are only ~0.3 mm. The brain consists largely of multipolar interneurons, meaning they have more than one input.

A neuron consists of a cell body (or soma), dendritic fibers, and an axonic fiber (see Figure 7.14). Just like in other cells, the cell body is tasked with keeping the cell alive and has a nucleus containing DNA. The neuron can have as many as 10,000 dendritic fibers that interconnect with other neurons and funnel input signals (i.e., electrical pulses) from other neurons to the cell body. When unstimulated the neuron has a slight negative voltage of –70 mV. However, if a sufficient number of electrical pulses are received to raise the cell to an activation potential of –55 mV, the neuron will fire a +70 mV pulse of ~1 millisecond duration down the axonic output fiber. The amplitude of a nerve impulse is fixed (see Figure 7.15). A stronger stimulus results in a proportionally higher number of pulses being sent down the axonic fiber. The axons are wrapped in myelin, a fatty substance that serves as insulation to keep the neurons from shorting out to each other. Breaks in the myelin sheaths, called the nodes of Ranvier, capacitively couple one segment of an axon fiber to the next, increasing the

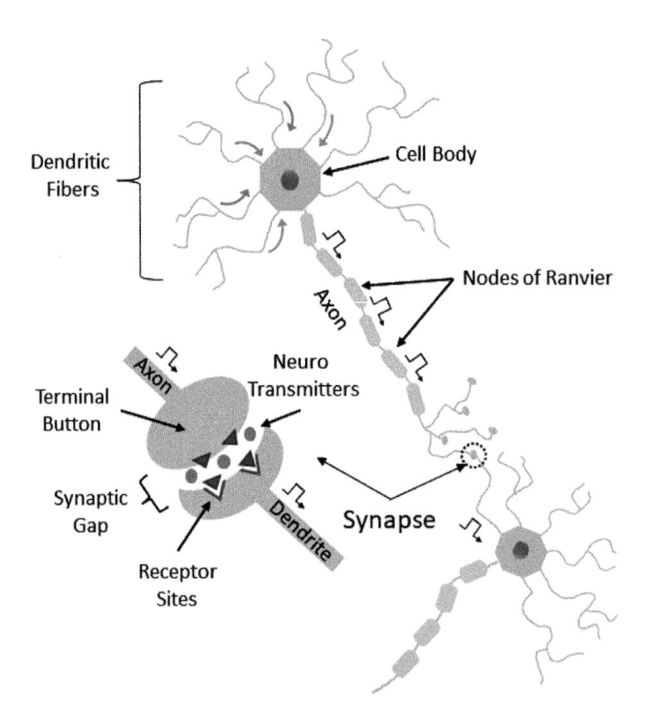

FIGURE 7.14 Neuron functional diagram. Neurons are specialized cells for processing information, which takes the form of electrical impulses. Like logic gates in a computer, neurons have either an "on" or "off" state. The neuron switches "on" if the voltage inputs from multiple dendritic fibers reach a threshold value (see Figure 7.15). When this happens the neuron generates an electric impulse that travels down the axon. The tip of the axon has multiple fibers which may overlap with the dendritic fibers of neighboring neurons. An incoming electrical impulse will trigger the release of chemical neurotransmitters that travel across the axonic fiber/dendritic fiber gap, referred to as a synapse, and provide a bridge for the pulse to travel to the neighboring neuron.

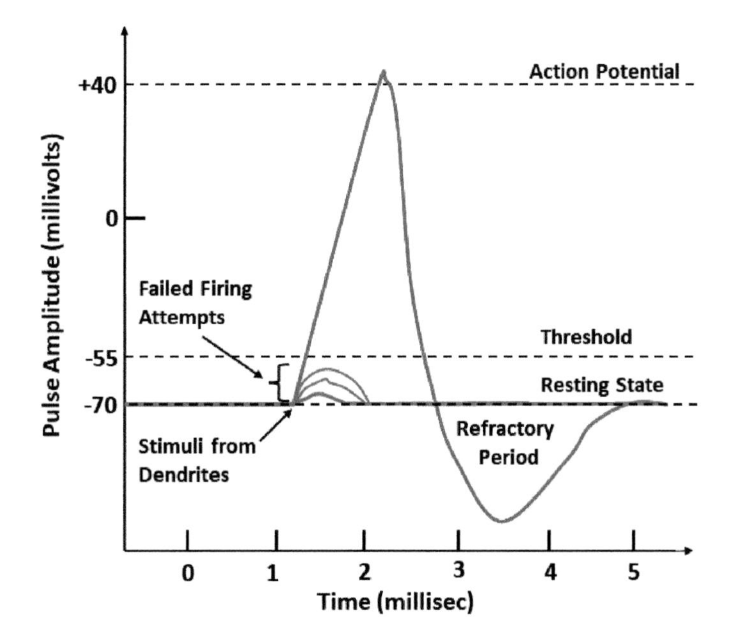

FIGURE 7.15 Neuron pulse. For a neuron to fire, the summed voltage presented to the cell body by the dendritic must exceed a threshold value.

propagation velocity of the nerve voltage pulse. The nerve synapse is the small gap about 30 nanometers wide occurring between the terminal button of the axon and the dendritic fiber of a neighboring neuron. If the pulse is sufficiently strong, its arrival at the terminal button triggers the release of chemical neurotransmitters that "swim" across the moat to receptor sites in the neighboring neuron's dendritic fiber. Receptor sites are "keyed" to accept specific neurotransmitters. When a neurotransmitter binds with a complementary receptor site it can either create a conductive bridge across which the electrical impulse from one neuron can travel to another or it can inhibit the formation of such a bridge, effectively stopping the pulse in its tracks. The more bridges that are created across a synapse, the stronger the pulse will be when it reaches the other side, and the more likely it will trigger the firing of a neighboring neuron. The travel of a pulse across a synapse is unidirectional, which makes it functionally analogous to a diode in an electrical circuit. After the passage of the pulse there is a built in ~1 millisecond delay (i.e., refractory period) to allow the neuron to return to its rest potential before another pulse can be processed. During this time unused neurotransmitters are reabsorbed, through a process called reuptake, into the terminal buttons. There are more than 100 types of neurotransmitters that can regulate everything from our appetite to our emotional state. Certain drugs can mimic naturally occurring neurotransmitters. For example, cocaine and morphine mimic dopamine, which triggers pleasure centers in the brain. Consumption of alcohol can both enhance the function of inhibitory neurotransmitters and decrease the function of excitatory neurotransmitters, resulting in an individual entering a drunken state with significantly reduced reflexes.

The more often a synaptic circuit is used, the stronger and longer lasting the synaptic bridges will be. These bridges represent the elementary building blocks of learning, memory, and cognitive activity. If we consider each synapse to be capable of storing 1 bit of information, and there are 10^4 dendritic fibers/synapses per 10^{11} neurons in the human brain, then the storage capacity of your brain is ~10^{15} bits, or equivalently, 1,000 terabits. A memory is not stored in a single synapse but is the consequence of a series of molecular, biochemical, cellular, and circuit-level changes in constellations of neurons throughout the brain (Langille and Brown 2018; Hebb 1949). Through memory the host is endowed with the ability to store knowledge and learn complex tasks which could either aid in its survival (like hunting and throwing a spear) or that make life more pleasurable (like writing music and playing an instrument). The more an individual performs such functions, the easier and more natural they will become, bringing a physical basis to the adage "practice makes perfect". In recent years, the storage capacity of our brains has been augmented by the availability of mobile devices that connect us to the internet. These devices have become part of our day-to-day existence. It is not hard to imagine that more efficient neural interfaces than the ability to type with one's thumbs will soon be developed. This point will mark an inflection point in our evolution, no less important than the evolution of the brain itself (see Figure 7.13).

The clock speed of a neuron is only ~ 1 kHz. The phenomenal processing power of the brain is due to its highly parallelized architecture, together with it having the features of a hybrid computer. A hybrid computer contains characteristics of both an analog and digital computer (see Figure 7.16). The digital part, with the on–off nature of a neuron, is efficient at performing logical and numerical operations. The analog part, which corresponds to the summing function of the dendritic fibers (see Figure 7.17), is an efficient solver of differential equations. Differential equations can be used to describe many physical phenomena and predict future events, such as figuring out the required initial velocity and trajectory for throwing a rock to hit a prey or how to move your body to catch an incoming ball. These capabilities are hardwired in our brains. This is why these actions come naturally, with

FIGURE 7.16 Photo of a Hybrid computer. A hybrid computer combines the lightning speed of an analog computer with the accuracy and control functions of a digital computer. The brain uses a similar mix of analog and digital approaches to process information. Hybrid computers were actively used in academia and defense until the 1990s when increases in the speed of digital computers reduced their competitive edge for many applications. The above is a 1966 photo of an Applied Dynamics AD/Four hybrid computer (Alamy).

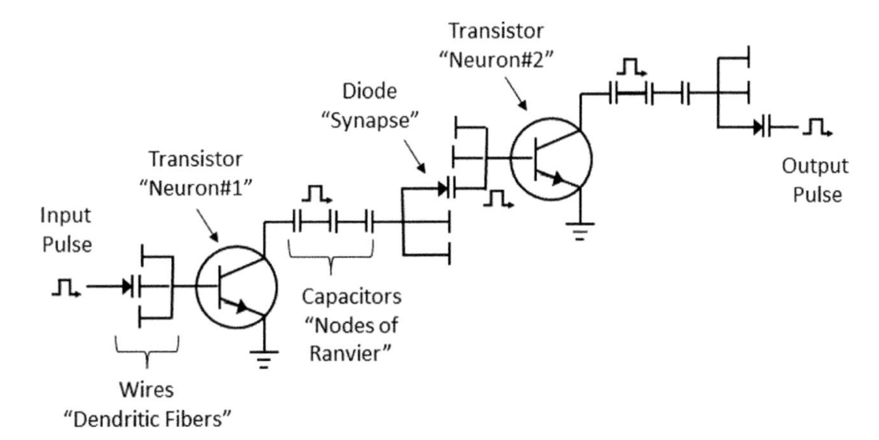

FIGURE 7.17 Neural network equivalent circuit. The above is a phenomenological equivalent circuit of Figure 7.14. Here semiconductor diodes replace electro-chemical synapses and transistors replace neuron cell bodies. A series of capacitors on the output of the transistors serve a similar function as the nodes of Ranvier on an axon.

little conscience effort. Unlike a digital computer, the speed of an analog computer is not determined by clock speed, but by the signal propagation speed, which can be a significant fraction of the speed of light.

7.6.3 Brain-to-Body Mass Ratios

The larger and more efficient the brain, the more complex the neural circuitry can be, suggesting that there may be a correlation between brain mass and intelligence. One might also expect that the larger the body of an organism, the larger the brain would need to be to control it. This leads to the hypothesis that a comparison of brain-to-body mass ratios could provide a means of comparing the intelligence between different species, with the implication that a greater brain-to-body mass ratio indicates greater intelligence. Another often used metric for comparing intelligence is the encephalization quotient, which compares actual brain mass to predicted brain mass based on body type across a range of species (Cairó 2011; Pontarotti 2016). As an example, primates have 5–10 times the brain mass expected for their body mass (Jerison 1973).

A brain-to-body mass ratio plot can provide valuable insights. In a seminal paper, Jerison (1969) charted brain:body data for 198 living vertebrate species on a log-log plot (see Figure 7.18). He included 94 species of mammals, 52 species of birds, 20 reptiles, and 32 fish. For any given species he chose the heaviest reported weights. His reasoning was that the collected animals likely represent more vulnerable, younger, or weaker individuals, and the heaviest example would be more representative of a given phenotype. When plotted, he found the data naturally separated the living vertebrate classes into two groups with respect to brain development; a "lower" vertebrate group, including fish and reptiles, and a "higher" vertebrate group, including birds and mammals. The plot shows that the brain-to-body mass ratios of mammals is, in general, ten to a hundred times that of contemporary reptiles. In the popular book "Dragons of Eden", Sagan (1977) added additional data points

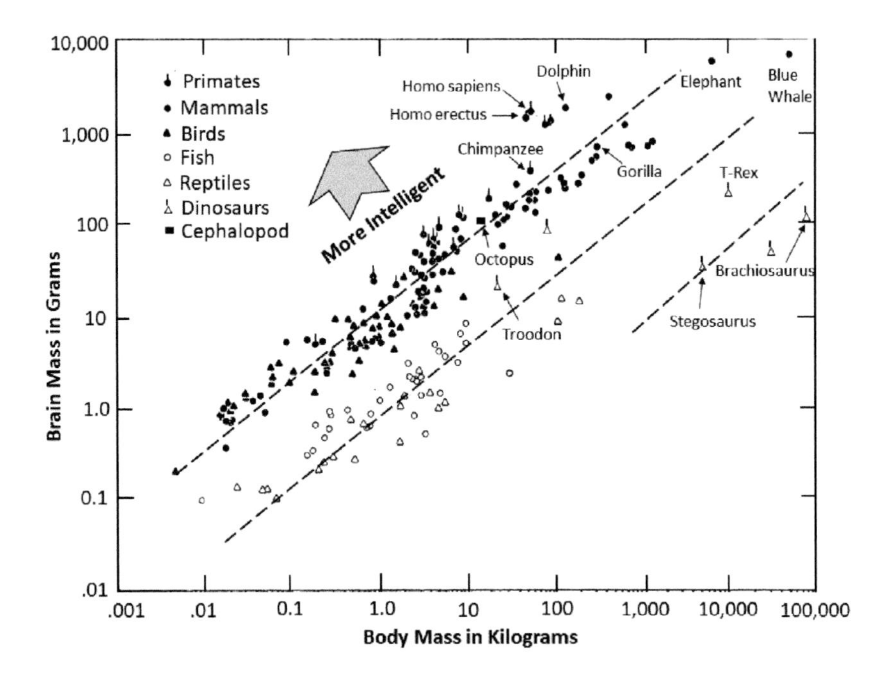

FIGURE 7.18 Brain-to-body mass ratios. Plots of brain-to-body mass ratios are often used to compare intelligence across species. The greater the ratio, the more intelligent a species is presumed to be. Under this assumption, the most intelligence species reside on the upper left-hand side of the plot. We reassuringly find ourselves in this position. Adapted from Jerrison 1969 and Sagan 1977.

for several extinct dinosaur and primate species. Although the data are sparse, inspection of the plot reveals extinct dinosaur species tend to have brain-to-body mass ratios an order of magnitude lower than contemporary reptile species.

If our hypothesis that greater brain-to-body mass ratios indicate greater intelligence is correct, then more intelligent species will fall on the upper left side of the plot. Somewhat reassuringly, we find that of those plotted, our species (*Homo sapiens*) is awarded the top spot under this intelligence paradigm. We also find that the species with the greatest brain-to-body mass ratios have only recently been added to life on Earth, suggesting that over time there is evolutionary pressure toward achieving greater intelligence. It also indicates that, if the evolution of life on planet Earth is typical, it takes billions of years for indigenous life to evolve the intelligence required to contemplate the existence of life elsewhere.

All vertebrate species possess the same basic brain structures: a forebrain, midbrain, and hindbrain. The forebrain is the forward-most part of the brain and is responsible for sensory processing, endocrine structures, and higher cognitive functions (e.g., forethought). The midbrain is involved in motor movement and visual/audio sensory processing. The hindbrain controls autonomic functions, such as blood pressure, respiration, and sleep cycles. As shown in Figure 7.19, this basic architecture has not changed over the course of time. However, the relative size and processing power of these structures have evolved over the past few hundred million years. In mammals the forebrain accounts for two-thirds of the total brain mass and is divided into left and right hemispheres. Each hemisphere is subdivided into frontal, parietal, temporal, and occipital lobes (see Figure 7.20 and 7.21).

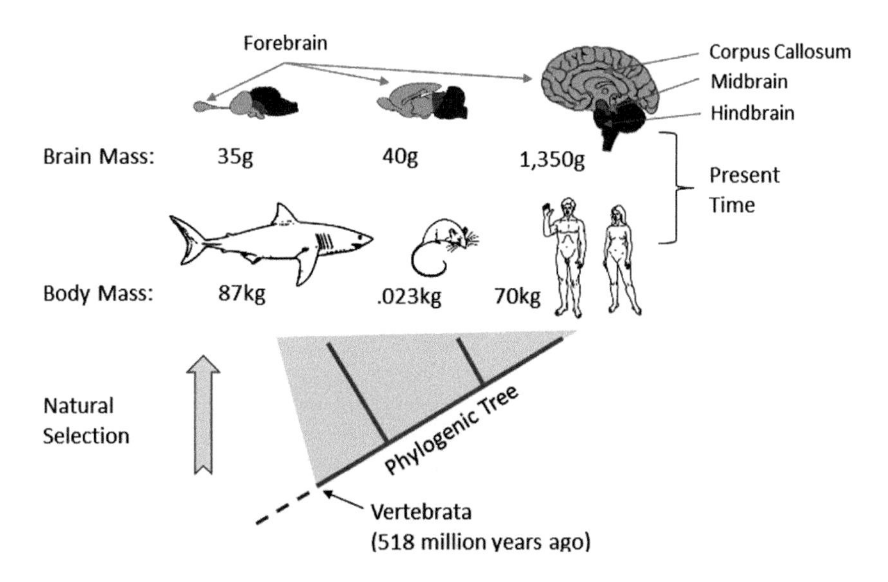

FIGURE 7.19 Vertebrate brain evolution. The brains of all vertebrates contain the same basic substructures; a forebrain, midbrain, and hindbrain. Of these structures, the forebrain shows the greatest volumetric variation across species. Adapted from Cesario et al. 2020.

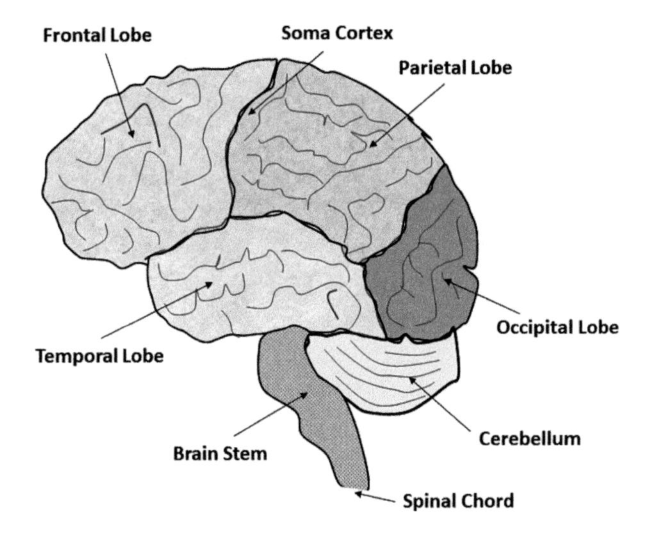

FIGURE 7.20 Brain lobes. The human brain is composed of multiple lobes, each responsible for different cognitive and/or bodily functions.

The frontal lobes are the seat of deliberation and regulation of action. The parietal lobe is the home of spatial coordination and the exchange of information with the rest of the body (via the soma cortex). The temporal lobe connects auditory and visual stimuli. Occipital lobes process visual information. The two hemispheres share data through the corpus callosum, a large C-shaped nerve fiber bundle running between the two hemispheres just below the cerebral cortex (see Figure 7.19 and 7.21). Interestingly, if the corpus callosum is severed, the higher brain functions of an individual are split, with very interesting consequences (see, for example, Coon and Mitterer 2010). The brain is power hungry, consuming

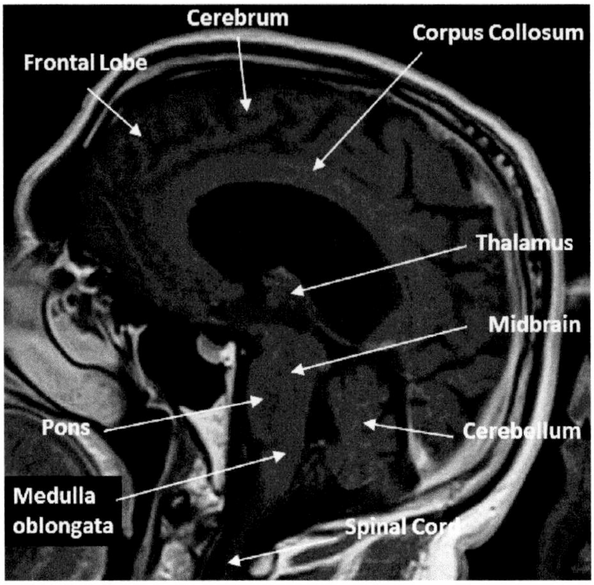

FIGURE 7.21 Brain cross section. Magnetic resonance imaging (MRI) can be used to identify features within a living brain.

20% of the body's total energy budget, despite representing only 2% of its mass. Intelligence comes at a relatively high metabolic cost (Herculano-Houzel 2012).

As a footnote to our discussion of brain-to-body mass ratios, let us consider the case of Laplace's brain (our hero from Chapter 5). Upon his death on March 5, 1827, at the age of 78, Laplace's physician, Francios Magendie, removed Laplace's brain and kept it on display in his home. The following is an account from a letter written by Joanna Baille to her great niece Miss Sophy Milligan about one such viewing (Pearson 1927). The letter, dated Hampstead (London), Monday, 1834, contains this notable paragraph:

"My dear Sophy … Dr. Somerville told us not long ago a whimsical circumstance regarding the head of La Place, the famous French astronomer. Some Ladies and Gentlemen went one day to the house of Majendie, the great anatomist, to see the brains of this philosopher which they conjectured must be of a very ample size, and seeing a preparation on the table answering their expectation they were quite delighted. "Ah! see what a superb brain, what organs, what developments! This accounts completely for all the astonishing power of his intellect, etc". Majendie, who was behind them and overheard all this, stepped quietly forward and said: "Yes, that is indeed a large brain, but it belonged to a poor Idiot, who when alive scarcely knew his right hand from his left. This, Ladies and Gentlemen" (handing to them a preparation of a remarkably small brain), "this is the brain of La Place". Dr. Somerville was told this anecdote by Majendie himself … Your affectionate Aunt, J. Baillie".

The average brain mass in humans is ~1350 gm, with a noticeable decline in cognitive abilities below ~850 gm. On the high end, Lord Byron (English poet) had a brain with a mass of 2,238 gm, while the brain of the Nobel prize winning French writer Anatole France came in at half that, 1,017 gm. Apparently, there can be large variation in brain size without significantly impacting the intelligence of its owner.

Sometime later, as the story goes, Laplace's brain became part of a traveling anatomical museum in England (Pearson 1929). This is a sad ending for Laplace, but fortunately his genius lives on in his published works.

7.6.4 Resident Alien

Careful examination of the brain-to-body mass plot of vertebrate species (Figure 7.18) will reveal that an "alien" intruder has been included, the octopus. The octopus is an invertebrate belonging to the Phylum Mollusca and Class Cephalopoda. Due to similar evolutionary pressures coming to bear, even phyletically distant cousins as cephalopods and mammals have developed complex neural systems with phenotypic similarities. This is an example of convergent evolution. Indeed, the development of a central nervous system has independently evolved more than five times on Earth. Examples include molluscs, annelids, nematodes, arthropods, and chordates (i.e., vertebrates) (Shigeno et al. 2018). As discussed by Shigeno, invertebrate nervous systems are highly diversified and often distributed, and do not resemble the more highly centralized nervous system of vertebrates. Of all invertebrates, the octopus has the highest brain-to-body mass ratio. From Figure 7.18 it can be seen to fall securely in the mammal range. The octopus can range in size from less than an inch in length to 100 pounds and 20 ft across (i.e., the Giant Pacific Octopus). The octopus brain contains ~500 million neurons, about the same as that of the family dog (Shuichi et al., 2018). A sketch of the nervous system of the octopus is provided in Figure 7.22. Approximately two-thirds of the neurons are in axial nerve cords that run from the primary brain down each arm. These are different from our spinal cord, in that

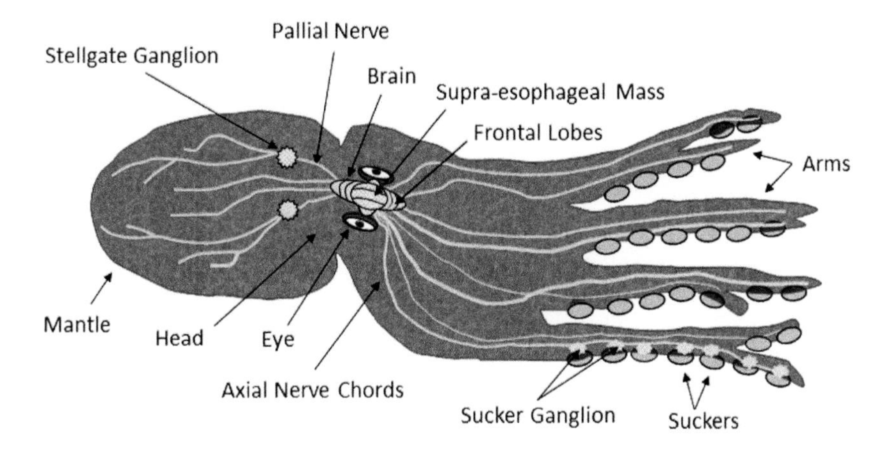

FIGURE 7.22 Resident alien. We are (or believe ourselves to be) the most intelligent of all the 65,000 vertebrate species currently living on Earth. Of the estimated 1.3 million invertebrate species, the octopus holds that honor. Its brain-to-body mass ratio puts it in the mammalian range (see Figure 7.18). Its brain mass is comparable to that of a dog and bestows it with significant cognitive abilities, including familiar behaviors. Unlike its mammalian cousins, it has a more distributed intelligence, with two-thirds of its neurons located outside the central brain. It is very different, but also similar to us in many ways. Given the apparently large number of planetary bodies with surface and/or subterranean oceans where a similar type of organism may thrive, this fact can bring us some comfort. Adapted from Figure 1: https://courses.lumenlearning.com/wm-biology2/chapter/diversity-of-nervous-systems/

they are not simple conveyors of impulses but appear capable of independent thought, with the primary brain serving as the "hive mind". Each sucker is connected to a nerve cord by a nerve ganglion, providing more sensory input than our finger tips, including taste. The octopus brain has more than 30 differentiated lobes. On either side of the brain are organs called statocysts containing a mass of sensitive hairs (akin to those in our inner ears) that allow the octopus to orient itself in three-dimensional space. The vision of an octopus is better than that of a human, with no blind spots. They are also able to detect the polarization of light, which may provide them with an uncanny ability to detect the motion of objects, by the virtue of being sensitive to subtle changes in reflected light.

Octopuses are capable of many complex behaviors, including maze and problem solving, indicating they have both short- and long-term memory (Zarella et al. 2015). They can be trained to differentiate between shapes, patterns, and even humans ... sometimes preferring the company of one over another. They have also been observed to play and, on occasion, go AWOL (absent without leave) from their aquariums in search of food and then return (Wood and Anderson 2004). The lifespan of an octopus is set by its reproductive cycle and can range from six months to as much as five years. After reproduction the octopus DNA orchestrates the organism's demise by preventing the repair and replacement of cells. This stage in the life of an octopus is called senescence and may last from weeks to a few months (Anderson et al. 2002). We humans suffer a similar fate in that there appears to be a limit to how often our DNA can faithfully replicate. Fortunately for us, our "self-destruct sequence" is not tied directly to reproduction, and we can persist for many decades, which allows us to accumulate knowledge and develop new capabilities. If it were not for this unfortunate feature in the DNA of an octopus, perhaps we would be preparing to explore the stars with our cephalopod cousins. Or, maybe, if they got a small evolutionary head start, they would let us tag along for the ride (see Figure 7.23).

FIGURE 7.23 Europans. From the perspective of Homo sapiens the octopus may seem alien. However, if our solar system is any indication, it may well be that moons and planets with subterranean oceans, like Europa, may greatly outnumber those with surface water. In such environs lifeforms akin to the octopus may evolve and reign supreme. Perhaps some day we will befriend such creatures on a distant world. [Image credit: masbt/flickr (CCO)]

7.7 SUMMARY

For much of Earth's history the most advanced form of life was single-celled organisms. Even now such organisms account for ~13% of the Earth's biomass. However, fortunately for us, a combination of astrophysical, geological, and chemical events, together with an inexorable evolutionary pressure, has led us to where we are today, on a planet with an estimated 8.7 million species of life, some of which are self-aware. Evolution has tailored life to survive and thrive in almost any environment; from tropical rainforests and glaciers to volcanic sea floor vents. The diversity and success of life on Earth gives us hope that we will find it elsewhere. In the next chapter, we will learn how to find other worlds and assess their suitability for life.

REVIEW QUESTIONS

1. What were conditions like on the primordial Earth?
2. What was the earliest/simplest lifeform on Earth?
3. What was the Oxygen Crisis?
4. What is punctuated equilibrium? Is there evidence for it?
5. How do random mutations occur?
6. Why are they a good thing?
7. What is the replication error rate in a virus?
8. What is the replication error rate in a eukaryotic cell?
9. When did sexual reproduction arise on the Earth?
10. What are some “Pros” of sexual reproduction?
11. What are some “Cons” of sexual reproduction?
12. In terms of biomass, which type of reproduction is the winner?
13. What is believed to have made the Cambrian Explosion possible?
14. When did the first organisms with photosynthesis arise on Earth?
15. How do we know when events happened in the distant past?
16. What have been the causes of mass extinction events in the past?
17. What is the likelihood of another one happening soon?
18. Name the types of neurons. Be able to label parts of a neuron cell.
19. What's a synapse?
20. How many neurons are in a typical human brain?
21. When did the first brains evolve on the Earth?

22. How many bits of info can be stored in the human brain?

23. How many bits of info can be stored in DNA?

24. What lifeform has dominated Earth the longest?

25. What is the typical size of the human brain in cm^3?

26. Below about what brain size does the individual start having cognitive problems?

27. What is the typical size of the human brain in cm^3?

28. Do humans have the largest brains on Earth?

29. Why do mammals rule Earth today?

30. What have been the major evolutionary leaps in the ability of life on Earth to store and utilize information?

31. What is the typical range in volume of the human brain?

32. What appears to be the best indicator of intelligence between species?

33. On what part of a brain-to-body mass ratio chart are the most intelligent species located?

REFERENCES

Aitken, M. J. 1990. *Science-Based Dating in Archaeology.* London: Longman.

Alroy, J. 2008. "Dynamics of Origination and Extinction in the Marine Fossil Record." *Proceedings of the National Academy of Sciences of the United States of America* 105 (Suppl. 1): 11536–42.

Alvarez, L. W., W. Alvarez, F. Asaro, and H. V. Michel. 1980. "Extraterrestrial Cause for the Cretaceous–Tertiary Extinction." *Science* 208 (4448): 1095–108.

Anderson, Roland C., James B. Wood, and Ruth A. Byrne. 2002. "Octopus Senescence: The Beginning of the End." *Journal of Applied Animal Welfare Science* 5 (4): 275–83.

Arnold, J. R., and W. F. Libby. 1949. "Age Determinations by Radiocarbon Content: Checks with Samples of Known Age." *Science* 110 (2869): 678–80.

Bell, Elizabeth A., Boehnike, Harrison Patrick, and T. Mark. 2015. "Potentially Biogenic Carbon Preserved in a 4.1 Billion-Year-Old Zircon." *Proceedings of the National Academy of Sciences of the United States of America* 112 (47): 14518–21. https://doi.org/10.1073/pnas.1517557112.

Berner, R. A. 1999. "Atmospheric Oxygen over Phanerozoic Time." *Proceedings of the National Academy of Sciences of the United States of America* 96, no. 20 (September): 10955–7.

Blackburn, Terrence J., Paul E. Olsen, Samuel A. Bowring, Noah M. McLean, Dennis V. Kent, John Puffer, Greg McHone, Troy Rasbury, and Mohammed Et-Touhami. 2013. "Zircon U-Pb Geochronology Links the End-Triassic Extinction with the Central Atlantic Magmatic Province." *Science.* 340 (6135): 941–5.

Boltwood, Bertram. 1907. "The Ultimate Disintegration Products of the Radio-Active Elements. Part II. The Disintegration Products of Uranium." *American Journal of Science* 23(Series 4), 77–88.

Buttefield, N. J. 2000. "Bangiomorpha pubescens: Implications for the Evolution of Sex, Multicellularity, and the Mesoproterozoic/Neoproterozoic Radiation of Eukaryotes." *Paleobiology* 26 (3): 386–404.

Cairó, O. 2011. "External Measures of Cognition." *Frontiers in Human Neuroscience* 5: 108.

Coon, Dennis, and John O. Mitterer. 22010. *Introduction to Psychology Gateways to Mind & Behavior*, Wadsworth Cenage Learning: Belmont, CA, p. 65.

Dempster, A. J. 1918. "A New Method of Positive Ray Analysis." *Physiological Reviews* 11, no. 4 (April): 316–25.

Dickin, Alan P. 2008. *Radiogenic Isotope Geology*. 2nd ed, 15–49. Cambridge: Cambridge University Press.

Dudley, Robert. 1998. "Atmospheric Oxygen, Giant Paleozoic Insects and the Evolution of Aerial Locomotor Performance." *Journal of Experimental Biology* 201, no. 8 (March): 1043–50.

Eldredge, N., and S. J. Gould. 1985. "Punctuated Equilibria: An Alternative to Phyletic Gradualism." In *Models in Paleobiology*, 82–115. San Francisco: Freeman Cooper. Reprinted in N. Eldredge. *Time Frames*, 193–223. Princeton: Princeton University Press.

Evans, N., 1996. "Extraterrestrial Life." *Burgess International Group Inc.*, Edina, Minn.

Hebb, D. O. 1949. *The Organization of Behavior*. New York: John Wiley and Sons.

Herculano-Houzel, S. 2012. "The Remarkable, Yet Not Extraordinary, Human Brain as a Scaled-Up Primate Brain and Its Associated Cost." *PNAS* 109 (Suppl. 1): 10661–8.

Jerison, H. 1973. *Evolution of the Brain and Intelligence*, 55–74. Academic Press: Cambridge, MA.

Langille, J. J., and R. Brown. 2018. "The Synaptic Theory of Memory: A Historical Survey and Reconciliation of Recent Opposition." *Frontiers in Systems Neuroscience* 26 (October).

Lobo, I. 2008. "Environmental Influences on Gene Expression." *Nature Education* 1 (1): 39.

Mani, K. 1996. "Permineralization Retrieved March 29, 2009, from Fossils: A Window to the Past." http://www.ucmp.berkeley.edu/paleo/fossils/permin.html.

Melott, A. L., B. S. Lieberman, C. M. Laird, L. D. Martin, M. V. Medvedev, B. C. Thomas, J. K. Cannizzo, N. Gehrels, and C. H. Jackman. 2004. "Did a Gamma-Ray Burst Initiate the Late Ordovician Mass Extinction?" *International Journal of Astrobiology* 3 (1): 55–61.

Mota, B., and S. Herculano-Houzel. 2014. "All brains are made of this: a fundamental building block of brain matter with matching neuronal and glial masses." *ront. Neuroanat.*, 12 November 2014. Volume 8: https://doi.org/10.3389/fnana.2014.00127.

Pearson, K. 1927. "The Brain of Laplace." *Nature* 119 (2998): 560.

Pearson, K. 1929. "Laplace, Being Extracts from Lectures Delivered by Karl Pearson." *Biometrika* 21 (December): 202–16.

Pennisi, E. 2012. "ENCODE Project Writes Eulogy for Junk DNA." *Science* 337 (6099): 1159–61. https://doi.org/10.1126/science.337.6099.1159.

Pontarotti, Pierre. 2016. *Evolutionary Biology: Convergent Evolution, Evolution of Complex Traits*, 74. Springer.

Ponting, C. P., and R. C. Hardison. 2011. "What Fraction of the Human Genome Is Functional?" *Genome Research* 21 (11): 1769–76.

Randall, L. 2015. *Dark Matter and the Dinosaurs*, 196–217. New York: Ecco/HarperCollins Publishers.

Rands, C. M., S. Meader, C. P. Ponting, and G. Lunter. 2014. "8.2% of the Human Genome Is Constrained: Variation in Rates of Turnover across Functional Element Classes in the Human Lineage." *PLOS Genetics* 10 (7).

Raup, D., and J., Sepkoski . 1982. "Mass Extinctions in the Marine Fossil Record." *Science* 4539(4539): 1501.

Ricci, J., X. Quidelleur, V. Pavlov, S. Orlov, A. Shatsillo, and V. Courtillot. 2013. "New 40Ar/39Ar and K–Ar Ages of the Viluy Traps (Eastern Siberia): Further Evidence for a Relationship with the Frasnian–Famennian Mass Extinction." *Palaeogeography, Palaeoclimatology, Palaeoecology*. https://doi.org/10.1016/j.palaeo.2013.06.020.

Sagan, C. 1977. *Dragons of Eden*. Random House Publishing Group: New York, New York.

Sawyer, S. A., J. Parsch, Z. Zhang, and D. L. Hartl. 2007. "Prevalence of Positive Selection among Nearly Neutral Amino Acid Replacements in Drosophila." *Proceedings of the National Academy of Sciences of the United States of America* 104 (16): 6504–10.

Sepkoski, J. J. 1984. "A Kinetic Model of Phanerozoic Taxonomic Diversity. III. Post-Paleozoic Families and Mass Extinctions." *Paleobiology* 10 (2): 246–67.

Shigeno, S., P., Andrews, G., Ponte, and G., Fiorito. 2018. "Cephalopod Brains: An Overview of Current Knowledge to Facilitate Comparison With Vertebrates." *Front Physiol* 2018 Jul 20: 9:952. doi: 10.3389/fphys.2018.00952. eCollection 2018.

Shuichi, S., P. Andrews, G. Ponte, and G. Fiorito. 2018. "Cephalopod Brains: An Overview of Current Knowledge to Facilitate Comparison with Vertebrates." *Frontiers of in Physiology*, Vol. 9, July 20, 2018.

Soddy, F. 1913. "The Radio Elements and the Periodic Law." *Chemistry News* 107: 97–9.

Sturtevant, H. 1913. "The Himalayan Rabbit Case, with Some Considerations on Multiple Allelomorphs." *American Naturalist* 47 (556): 234–8.

The ENCODE Project Consortium. 2012. "An Integrated Encyclopedia of DNA Elements in the Human Genome." *Nature* 489 (7414): 57–74.

von Frese, R., L. Potts, S. Wells, T. Leftwich, and H. Kim. 2009. "GRACE Gravity Evidence for an Impact Basin in Wilkes Land, Antarctica." *Geochemistry, Geophysics, Geosystems* 10 (2). https://doi.org/10.1029/2008GC002149.

Whitman, W. B., D. C. Coleman, and W. J. Wiebe. 1998. "Prokaryotes: The Unseen Majority." *Proceedings of the National Academy of Sciences of the United States of America* 95 (12): 6578–83. https://doi.org/10.1073/pnas.95.12.6578.

Wilson, A. T., and D. J. Donahue. 1992. "AMS Radiocarbon Dating of Ice: Validity in the Technique and the Problem of Cosmogenic In-Situ Production in Polar Ice." *Radiocaron* 34 (3): 431–5.

Wood, J. B., and R. C. Anderson. 2004. "Interspecific Evaluation of Octopus Escape Behavior." *Journal of Applied Animal Welfare Science* 7 (2): 95–106.

Zarrella, I., G. Ponte, E. Baldascino, and G. Fiorito. 2015. "Learning and Memory in Octopus vulgaris: A Case of Biological Plasticity." *Current Opinion in Neurobiology* 35: 74–9.

CHAPTER 8

Habitable Worlds

PROLOGUE

From observations of Cepheid variable stars, the diameter of the Milky Way is estimated to be 258,000 light years. From the Doppler shifts observed in stellar spectra, we know it has a rotational velocity of ~220 km/s. This information, together with Kepler's and Newton's Laws, can be used to estimate the mass of our galaxy. With just a few taps on a phone calculator, we find the mass of our Milky Way to be ~ 1.5 *trillion* solar masses. About 4% of this mass is in the form of stars, 12% is in the form of gas, and the remaining 84% is in dark matter. An average star has a mass of about one-quarter that of our Sun. An estimate for the number of stars in the Milky Way is then 1.5 trillion stars × 0.04 ÷ 0.25 = 240 billion. Planets are a natural byproduct of the star formation process. Our solar system has eight planets. If this is typical, then the total number of planets in our galaxy is something like 8 × 240 billion, or ~ 2 trillion planets. How many of these will be capable of supporting life? The answer to this question depends on many factors. These include the spectral type of the host star, where the planets end up orbiting the host star, and whether or not the planet is rotating and has an atmosphere. Assuming a planet is fortunate enough to spawn life, how would we ever know? The nearest known "Earth-like" planet (Proxima Centauri b) is 4.24 light years distant. The fastest spacecraft is Voyager 1, travelling at 38,000 miles per hour. If it were headed toward Proxima b (which it is not), it would take about ~75,000 years to get there and another 4.24 years to radio back its observations, making this approach to assessing habitability currently intractable to us. In this chapter, we will discuss approaches for identifying habitable worlds and worlds with life utilizing technologies that are now within our grasp.

8.1 INTRODUCTION

So, what does a planet need to support life? We know that the presence of liquid water is essential to life on Earth. Liquids provide a medium through which molecules and atoms can interact. It is also essential to the formation and destruction of the peptide bonds that hold proteins together (see Chapter 5). Without water, the bonds are broken, and we literally turn to dust. Usually, we think of temperature as being the only factor in determining

DOI: 10.1201/9781315210643-8

if water will be in a solid, liquid, or gas phase. But atmospheric pressure is also important. At sea level water exists in a liquid state between about 273 and 373 K, or, equivalently, 0 and 100°C. At lower temperatures, its solid and organic chemistry is "frozen" in time, unable to make and break chemical bonds. (This is why freezers keep food from spoiling.) The lower the pressure, the lower the temperature at which a liquid will boil. (This is why there are high-altitude cooking instructions.) The higher the pressure, the higher the temperature at which a liquid will boil. If the pressure is really low (<0.006 atmospheres), water will not become a liquid at any temperature, but will sublimate, i.e., go directly from a solid to a gas.

8.2 HABITABLE ZONES

Living cells have been found in hot springs close to the boiling point of water (373 K). Similarly, life has been found at near-freezing temperatures (273 K). Let's take this range of temperatures, 273 K–373 K, as the one suitable for life. Therefore, one of the key requirements for the origin and evolution of life on the surface of a planet is that the planet has an atmosphere and surface temperature commiserate with the presence of liquid water.

An airless planet at a distance, d, from a star with a luminosity, L, will have an equatorial surface temperature given by (Huang 1959; Evans 1996),

$$T = 393\left(\frac{(1-A)L}{d^2}\right)^{\frac{1}{4}} \tag{8.1}$$

where

T = temperature on surface of planer (K)
L = luminosity of star in units of solar luminosity ($L_\odot$)
d = distance from the star in astronomical units (AU)
A = planetary albedo (i.e., the fraction of energy reflected by the planet)

The above expression tells us that to double the temperature on the surface of an airless world, like the Moon, would require the luminosity of its star to increase 16 times (!), which makes the surface temperature less vulnerable to minor variations in the Sun's output. Stars on the main sequence (like our Sun) are in hydrostatic equilibrium (see Chapter 4). If its core compresses, then the nuclear reaction rates go up and central temperatures go up. The star responds by expanding and cooling back toward the original surface temperature. If the reaction rate slows and the core cools, then the star will contract and heat back up. This thermostatic action has kept the energy output of the Sun stable to within 0.2% over the past few million years, which is great for providing a nurturing environment for the evolution of lifeforms such as ourselves.

The albedo, A, is the fraction of star's energy reflected by the planet back into space. The fraction of energy absorbed by the planet is then (1–A). As an example, the albedo of snow is about 1, indicating it reflects almost all the light that hits it. The albedo of charcoal is only 0.04, meaning it absorbs 96% of incident light. That is why it appears so black. The albedo of the Moon is just 0.12 (about the same as gravel). It only looks white because the Sun is so bright. The value of both L and d for the Moon is the same as for Earth, 1.0 (by definition).

FIGURE 8.1 Goldilocks zone. The range of distances from a star over which conditions are "just right" for liquid water to exist on a planet's surface, i.e., "not too hot and not too cold". Image Credit: Shutterstock.

Substitution into Eq. (8.1) then yields a temperature for the lunar surface of 381 K (astronaut boots need to be insulated!). Another factor that can influence the temperature of a planet is whether or not it is rotating. When the daytime surface of a planet swings around to the nightside, it will radiate some fraction of the energy it absorbed during the day back into space, lowering the average surface temperature. How much the planetary surface will cool down depends on the atmospheric composition. A planet with a high concentration of greenhouse gases, like Venus, will trap heat and cool down little. The Earth's albedo is ~0.4. Substitution into Eq. (8.1) yields an equatorial surface temperature of ~346 K or 163°F, which is then moderated by day–night variations. One of the highest recorded temperatures on Earth is 177°F (80.8°C) in the Lut Desert of Iran (Zhao, et. al. 2021).

A star's habitable zone (*aka*, Goldilocks zone or ecosphere) is the range of orbital radii over which surface conditions on a planet may be conducive to the existence of liquid water (see Figures 8.1 and 8.2). The term habitable zone was coined by Su-Shu Huang in 1959 (Huang 1959), but it was first discussed by none other than Isaac Newton himself some 272 years earlier in his "Principia" (Newton 1687). Quoting Newton (Lingam 2021):

> Our water, if the earth were located in the orbit of Saturn, would be frozen, if in the orbit of Mercury, it would depart at once into vapours. For the light of the sun, to which the heat is proportional, is seven times denser in the orbit of Mercury than with us: and with a thermometer I have found that with a seven-fold increase in the heat of the summer sun, water boils off.

The inner boundary of the zone, d_{inner}, is where the planet is so close to the star that for any reasonable atmospheric pressure water on the surface would evaporate. Likewise, the outer

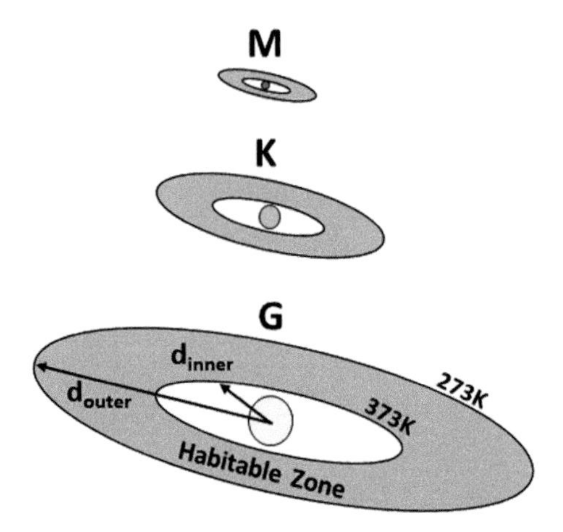

FIGURE 8.2 Habitable zones. The habitable zone is the range of orbital radii over which a planet has the best chance for liquid water to exist on its surface. The width of a habitable zone depends on the luminosity (i.e., spectral type; *e.g.*, M, K, or G) of the parent star.

boundary is the distance, d_{outer}, at which surface water would freeze. d_{outer} is the same as the snowline discussed in Chapter 5. There values depend on the luminosity of the host star. Mathematical expressions for d_{inner} and d_{outer} can be derived from the Stefan–Boltzmann Law Eq. (4.4).

$$d(AU) = 1.55 \times 10^5 \left(\frac{L(L_\odot)}{T^4} \right)^{\frac{1}{2}} \tag{8.2}$$

By substituting in the boiling and freezing points of water, expressions for d_{inner} and d_{outer} can be derived.

$$\begin{aligned} d_{inner} &= 1.1\, L^{\frac{1}{2}} \\ d_{outer} &= 2.1\, L^{\frac{1}{2}} \end{aligned} \tag{8.3}$$

Substitution of $L = 1$ for our Sun yields current values of $d_{inner} = 1.1$ AU and $d_{outer} = 2.1$ AU. This simple derivation only serves as a rough estimate of the location and extent of habitable zones and does not take into account the effects of planetary atmospheres discussed above or the possibility of subsurface life.

Comparing Eqs. (8.1) and (8.2), we can see that the boundaries of the habitable zone are more sensitive to changes in stellar luminosity than planetary surface temperature. Indeed, since arriving on the main sequence the Sun's luminosity has slowly risen from ~70% of its current value. This means early on the habitable zone was shifted closer to the Sun, making Venus more habitable. As the Sun's luminosity increased it pushed d_{inner} out well beyond Venus, making the origin and evolution of life there more problematic. The Sun will continue to slowly grow more luminous over the next few billion years. Ultimately, the Earth will suffer the same fate as Venus, pushing humanity to Mars, the outer solar system, and beyond (see Figure 8.4).

Our species took ~4.7 billion years to evolve on the Earth. Assuming this is the norm, for a star to be considered a "good star" for hosting the evolution of intelligence, it should have a lifetime, t_{life}, similar to that of our Sun, ~10 billion years. The more luminous a star is, the faster it burns through its supply of fuel. For main sequence stars, i.e., stars burning hydrogen into helium in their cores, the following expression can be used.

$$t_{life} \approx \frac{1}{L^{0.833}} \tag{8.4}$$

where

t_{life}= lifetime of star in units of solar lifetimes (10 billion years)
L = luminosity of star in units of solar luminosity ($L_{\odot}$)

Using the above expression, a spectral type K main sequence star with a luminosity of 0.5 $L_{\odot}$ would be expected to live 1.78 times as long as the Sun, or 17.8 billion years. A type F star, with twice the luminosity of the Sun (a class G star), would last about 5.6 billion years. Likewise, the mass of a main sequence star can be estimated using the relation (see Chapter 4),

$$M \approx L^{\frac{1}{3}} \tag{8.5}$$

where

M = mass of star in units of solar masses ($M_{\odot}$)
L = luminosity of star in units of solar luminosity ($L_{\odot}$)

Therefore, a star with a luminosity, L, of 0.5 $L_{\odot}$, would have a mass of 0.79 $M_{\odot}$ and a star with a luminosity twice that of the Sun would be 1.26 $M_{\odot}$.

A high luminosity star will harbor a wide habitable zone that extends out to large distances. This may increase the chances that planets may reside in it. However, the star will likely not survive long enough for life to have a chance to evolve. A low-luminosity star will have a narrower habitable zone closer in to the parent star. The small width of the zone could make it less likely planets will be found there. Figure 8.3a is an H-R diagram of stars in our solar neighborhood. Plots of habitable zone width and stellar lifetimes of a subset of these stars are plotted as a function of spectral type in Figure 8.3b. From these results, it would appear the range of mass and luminosity over which a star is good for harboring a planet suitable for the origin and evolution of intelligent life is fairly narrow. Perhaps not unexpectedly, our Sun is in the middle of it. These are the stars, spectral type F through K, that are likely the best candidates for hosting an intelligent civilization.

8.3 FINDING PLANETS

The idea of planets orbiting stars other than our own is nothing new. The possibility of exoplanets was widely considered by Aristarchus and others (see Chapter 2) over 2,000 years ago and revisited by many investigators over the intervening millennia. The problem in proving their existence has been distance. Humans have systematically underestimated

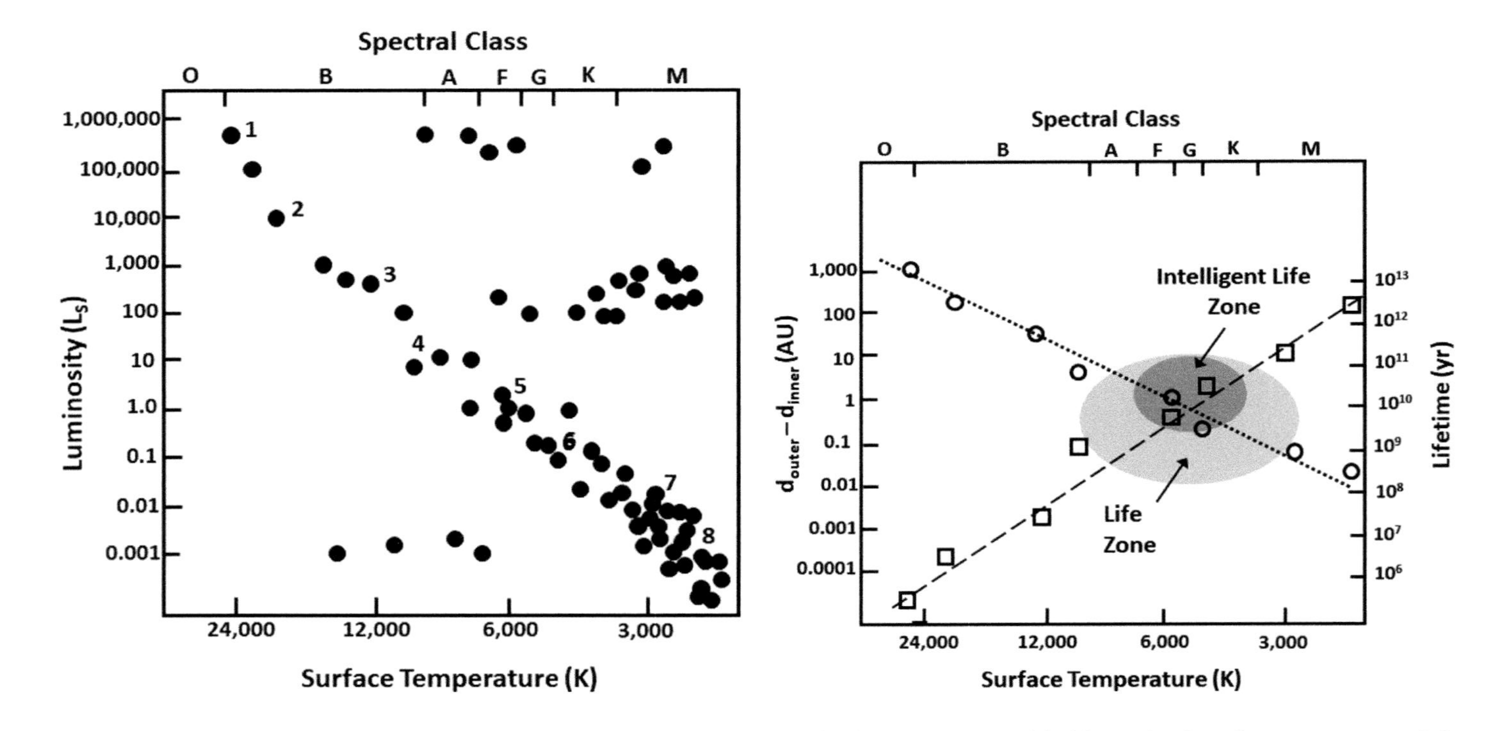

FIGURE 8.3 Best candidate stars. (a) H-R diagram of nearby stars. (b) Plots of habitable zone width (d_{inner}–d_{outer}) and main sequence lifetime as a function of spectral type. In general, the wider the habitable zone the greater the likelihood of finding planets within it. The greater the stellar lifetime, the more time for the evolution of life. The cross-over point in the plot indicates the spectral class with the optimum balance between these two desirable characteristics. Perhaps not unexpectedly, our Sun (a G star) is right there. Life originated on Earth within the first billion years or so after it formed. If this is typical, then stars must live at least this long for life to evolve on their worlds (Life Zone; spectral class A through M). The evolution of intelligent life demands stellar lifetimes of several billions of years (Intelligent Life Zone; spectral Class F through K).

the distance to astronomical objects. Each discovery pushed things out that much farther. Detecting an exoplanet was something like trying to catch a rainbow. There are numerous techniques that have been proposed to detect planets around other stars. Some have been much more successful than others.

8.3.1 Direct Imaging

Once a solar system forms, a star holds onto its planets like a protective parent, keeping them from the perils of interstellar space. (However, some wayward offspring have been found, see for example, Section 8.3.4.) The distance between stars is great. The closest star, Proxima Centauri, is 4.26 light years, about 25 trillion miles away. At that distance a planet with an Earth-type orbit extends, at most, only 0.7 arcseconds from the star. The angular resolution of the human eye is at best ~60 arcseconds. A telescope with just a 10-inch (~25 cm) lens or mirror could, in principle, resolve a planet around Proxima Centauri. Such telescopes were available by the eighteenth century, but their images were blurred by atmospheric turbulence (i.e., seeing). This can be mitigated to some degree by locating telescopes on high mountaintops, above as much of the troublesome atmosphere as possible. There is also the vexing problem that planets shine by reflected light from their Sun (like the Moon does), with the result being that their feeble light is lost in the glare of their stellar parent.

The issues associated with directly imaging a nearby planetary system were first overcome in 2004 with the observation of 2M1207b, a gas giant planet five times more massive than Jupiter in orbit about a brown dwarf, 2M1207, at a distance of 172 light years (Chauvin et al. 2004). A brown dwarf is a substellar object (in this case ~25 Jupiter masses) whose central temperature is not high enough to sustain a nuclear fusion reaction. 2M1207b was observed in the near infrared, where the planet is just 100 times fainter than 2M1207 (see Figure 8.4). For comparison, at optical wavelengths, the

FIGURE 8.4 First Exoplanet Image. 2M1207b is a five Jupiter mass planet (orange circle) orbiting a brown dwarf. It was first observed in 2004 using the Very Large Telescope (VLT) in Chile. Image Credit: ESO.

brightness ratio between an Earth-like planet and a star like the Sun is a billion to one. The observation was made from the Very Large Telescope (VLT) located in the high Andes of Chile using an adaptive optics (AO) system. AO systems utilize fast-moving mirrors to compensate for turbulent motions in the Earth's atmosphere. In addition to AO, coronagraph systems have been developed to reduce the glare of the central star by occulting it with a disk. (You may have done something similar at a nighttime outdoor event, where holding up your thumb against stadium lights reveals a swarm of moths.) While the most informative images come from planetary systems seen face-on, they can also be used to probe edge-on systems. Since 2004 more than two dozen planetary systems have been directly imaged. This method accounts for ~60 exoplanet detections (only ~1% of the total) and works best for observing massive planets at infrared wavelengths far from the parent star. Indirect methods of detecting exoplanets that do not rely on high angular resolution have proven to be far more successful, but perhaps a bit less satisfying to our image-oriented brains.

8.3.2 Doppler Spectroscopy

The first detection of a planet orbiting a main sequence star was achieved not through direct imaging, but by Doppler spectroscopy. As we learned in Chapter 3, the first photograph of absorption lines in the solar spectrum was made by John Draper in 1843 using a telescope and early camera (Figure 2.12). By 1868 the technology had improved sufficiently that absorption line spectra could be taken toward distant stars. One of the first such observations was made by William Huggins toward the star Vega. He noted that the wavelengths at which the absorption lines occurred were shifted from where they occurred in laboratory spectra. He concluded that the shift was due to the star moving at a velocity of 32 km/s relative to Earth (Section 2.5). The shift in wavelength or, equivalently, frequency due to the motion of an object is referred to as the Doppler Effect. The use of such shifts to deduce the motion of objects is called Doppler spectroscopy. In 1952, Otto Struve pointed out that the gravitational pull of a sufficiently massive planet, such as Jupiter, would cause a star to move back and forth relative to an observer on Earth. He proposed that a sufficiently sensitive spectrograph could pick up such oscillatory motions toward stars and, with the help of Kepler's and Newton's Laws, used to infer the presence, number, orbital radius, and mass of planets (Struve 1952; see Figure 8.5). Forty years later spectrograph technology had advanced to the point where observers could detect velocity shifts as low as 7 m/s. Michel Mayor and Didier Queloz used such a spectrograph on a 1.3 m telescope to detect a Jupiter size planet orbiting close to the main sequence star 51 Pegasus, a Sun-like star ~50 light years from Earth (Mayor and Queloz 1995). The detection of such a planet, 51 Pegasi b (referred to as a "hot Jupiter"), so close to a star was completely unexpected and forced a rethinking of how planetary systems evolve. This is one of many examples of observation-leading theory. Since the amount of observed Doppler shift is directly proportional to the mass of the planet, the inverse square of its orbital distance from the star, and the cosine of the angle of inclination to the observer, Doppler spectroscopy works best for edge-on systems composed of massive planets orbiting close to the star. Face-on systems

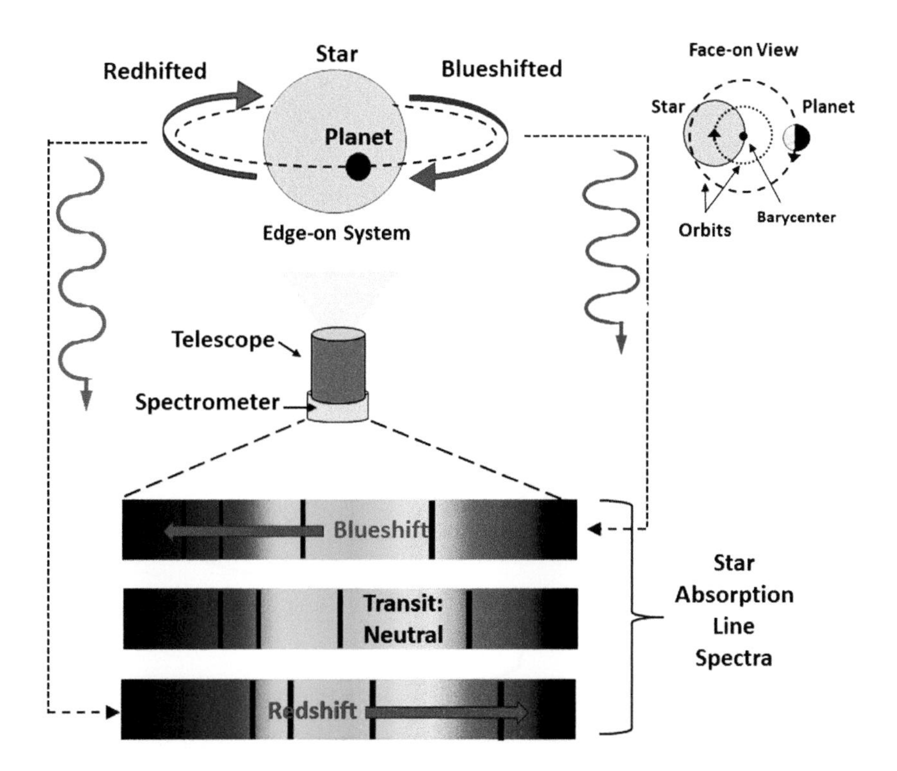

FIGURE 8.5 Doppler spectroscopy. The center of mass of two orbiting bodies is located between them. This is the point about which the two bodies orbit, referred to as the barycenter. If the body masses are greatly unequal, it may be located within the more massive of the two. This can be the case for planets orbiting stars. For example, the center of mass of our own solar system is located near the surface of the Sun. The orbital motion about the barycenter makes the star wobble back and forth over time. The velocity associated with the wobble causes the star's absorption line spectrum to be Doppler shifted. When the wobble pulls the star away from us, it stretches out the spectrum causing it to appear redshifted. Likewise, when the wobble pulls the star toward us, the spectrum is compressed, causing it to be blueshift. The more massive the planet and the closer it is to the star, the greater the shifts will be and the easier they will be to detect. Due to velocity projection effects, the Doppler shifts are greatest when the planetary system is observed edge-on, with face-on systems showing no shift.

are undetectable with this approach. Since this first observation, ~1,000 extrasolar planets (~20% of the total) have been detected in this way. A listing of nearby worlds found using Doppler spectroscopy is provided in Appendix A2.

8.3.3 Planetary Transits

In planetary systems that are nearly edge-on to us, planets will occasionally cut-in front of their star, blocking some of the star's light from reaching an observer. This stellar occultation by the planet will occur once per orbit. Therefore, by precisely monitoring the dimming of a star's light over time (and judiciously applying Kepler and Newton's Laws), it is possible to determine the number, orbital radius, and size of orbiting planets (see Figure 8.6). In the case of a Jupiter size planet orbiting a Sun-like star, such as HD209458, the passage

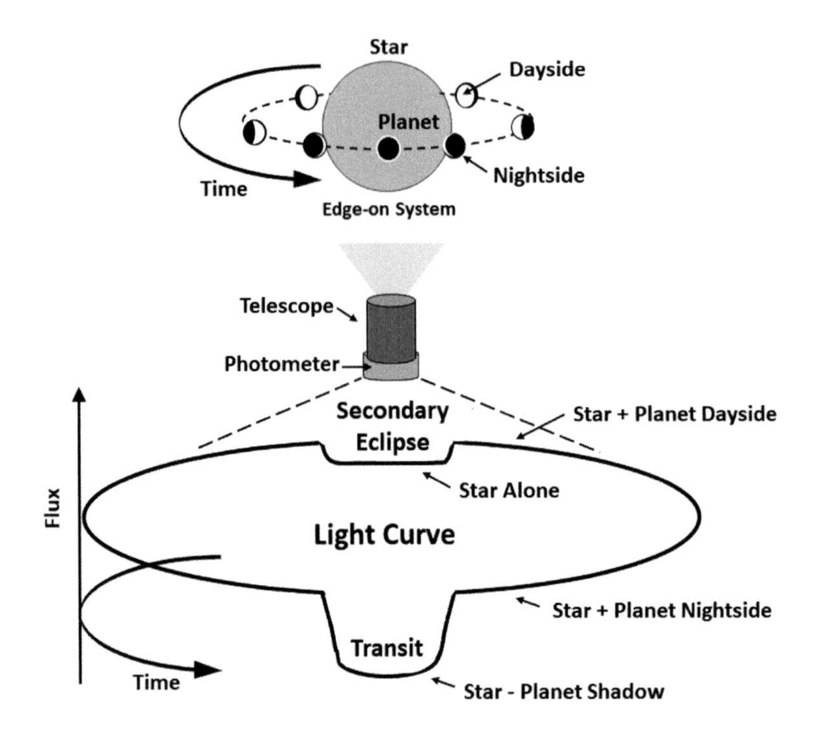

FIGURE 8.6 Transit technique. If a planetary system is edge-on to us, then during the course of their orbits the planets will pass in front of their star, blocking a small fraction of the star's light, making it appear dimmer. This occurrence is referred to as a transit. Therefore, through carefully monitoring the light curve of a star over multiple transits it is possible to determine the presence, size, and orbital characteristics of a planet; including whether it is in a star's habitable zone. Used with permission from Joshua N. Winn.

of such a planet can reduce the star's brightness by as much as 1.7% (Charbonneau et al. 2000). The larger the diameter of the planet's silhouette relative to the star's diameter the stronger the transit signal will be. For terrestrial planets orbiting a Sun-like star the reduction in brightness is only about 0.01%. Given these small brightness variations, the detection of planets using the transit technique is best done from space, free from the deleterious effects of atmospheric turbulence. However, there are many examples of successful ground-based transit programs (e.g., SuperWASP, the MEarth Project, and KELT). Since the transit technique only requires the accurate measurement of light (i.e., photometry) of a star and not a complex spectrograph, it can be performed using a camera with a field of view large enough to observe more than 100,000 stars at once, making large planetary surveys possible. This more than compensates for the fact that the fraction of planetary systems that are sufficiently edge-on is expected to be less than 0.5%. Such a large survey was performed by the 0.95 m *Kepler* space telescope launched in 2009 (Borucki et al. 2010). Over its 9-year mission, *Kepler* observed 530,506 stars and detected 2,662 planets. Such large surveys where multiple stars are observed allow statistical studies to be performed, reminiscent of the early spectroscopic surveys performed by Draper, Higgins, and Pickering that ultimately led to our understanding of stellar evolution (see Chapter 3). Systems with single planets can have a 40% false detection rate. Eclipsing binary systems can also yield false planet detections.

However, follow-up observations using Doppler spectroscopy and transit-timing variations (TTVs) techniques can be used to confirm the presence of planets and estimate their mass. TTVs occur in close-packed, multi-planet systems where gravitational interactions between planetary members cause some planets to speed up and others to slow down, changing the timing between transits from what would be expected if only one planet was present. As we will soon discuss, the transit technique can also be used to probe the atmospheres of exoplanets. To date, of the ~5,000 exoplanets that have been discovered, ~80% of the detections were performed using the transit technique. Many more will follow.

8.3.4 Astrometry

The astrometric technique for detecting planets is akin to the Doppler spectroscopy approach. When we look up at the night sky, the stars appear to be fixed relative to one another. But if we were to observe the sky over many years, we would see they move. These movements are referred to as proper motions. Stars closer to us have more noticeable proper motion than more distant ones; just like a jet plane that is closer to us appears to move faster than one further away, even though the one further way could be going faster. Our far less distracted ancestors took note of such things, partly because the lack of artificial light and pollution made the stars appear brighter against the black of space, and also because the locations of the constellations were routinely used to mark the passing of the seasons. Movements of planets or the occurrence of comets were also tracked for their religious significance (see Chapter 2). If a star with a noticeable proper motion has a planet with an orbital plane perpendicular to our line of sight, the gravitational pull of the planet will cause the star to wobble back and forth in a sinusoidal pattern as it moves across the sky (see Figure 8.5). The more massive and/or closer the planet is to the star, the greater the wobble will be. Its presence could, in principle, be revealed by taking time-lapse images of the star. Each orbit of the planet would cause the star to complete one wobble.

Since imaging is required, this approach to planet detection suffers from many of the same drawbacks as the direct imaging technique. The first evidence for a planet being detected using astrometry was reported in 1855 (Jacob 1855), and there have been several since, but none appear to have stood the test of time. To date, neither ground- nor space-based observations have yielded the discovery of an exoplanet using astrometry. However, an astrometric wobble due to the presence of a previously discovered planet orbiting Gliese 86 was observed by the Hubble Space Telescope (Benedict 2002).

8.3.5 Gravitational Microlensing

One prediction of Einstein's Theory of General Relativity is that stars warp the space around them. How much the space is warped depends on the mass of the star. This prediction was confirmed by Arthur Eddington (one of our story's heroes from Chapter 4) by observing the bending of starlight around the Sun during a total solar eclipse in 1919. The warped region of space acts like a gravitational lens, focusing the light from background objects. Let us consider the case of two stars and an observer positioned along a straight line, with the observer located at the focus of the gravitational lens produced by the middle star (see Figure 8.7). Light from the background star (the one at the top of the figure) is bent toward

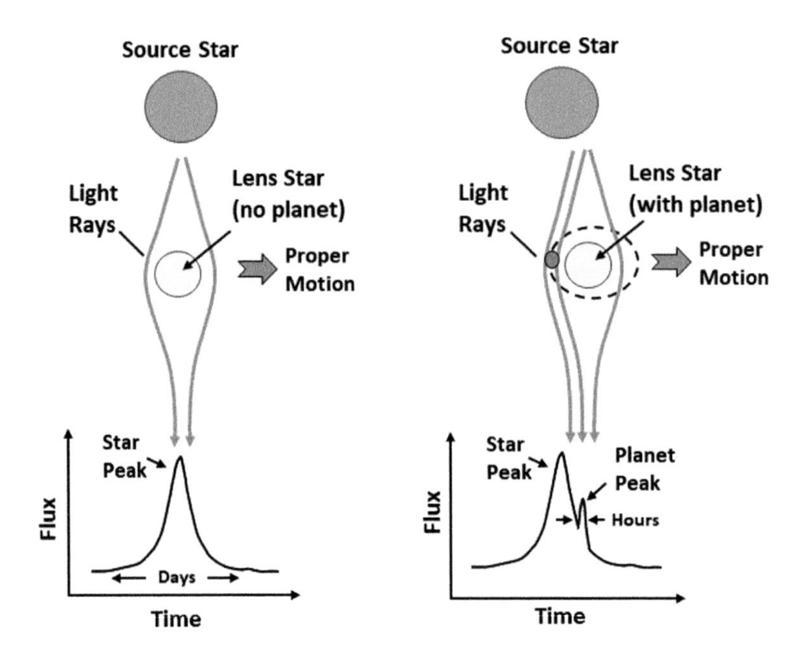

FIGURE 8.7 Gravitational microlensing. Stars warp the fabric of space around them. Gravity is a physical manifestation of this warp. The paths of photons passing through these regions of warped space will be bent as if the photons were passing through a lens. If a light detector is located at the focus of the gravitational lens, the light from a star located behind the lens star will be greatly amplified and added to the light (i.e., flux) measured in the direction of the lens star. The amplification will only last as long as the light detector/ observer and lens star are in near perfect alignment. Due to the relative motion of the source star, lens star, and observer such an alignment will typically last for only a few days, yielding a peak in the light curve. If the lens star has one or more planets, it will produce secondary peaks in the light curve. The brighter the secondary peaks, the greater is the mass of the planets. Therefore, by carefully monitoring the timing and intensity of these peaks, it is possible to discover and probe very distant planetary systems.

the observer, adding to the brightness of the foreground star that is creating the gravitational lens. When this occurs the foreground star will appear orders of magnitude brighter to the observer than it otherwise would. Since the background and foreground stars must be in a straight line for this to occur, their relative motion will make such events transitory, lasting only weeks or months before the alignment is broken. The observer would see a large peak in the star's apparent brightness during the alignment. If the foreground star happens to have a planet, then as the planet swings around the star, its gravity can act as an additional lens and further amplify the light from the background star, leading to the appearance of a secondary peak in the light curve. How long the secondary peak lasts will depend on the planet's orbital speed, which, we know from Kepler's Laws of planetary motion, is a function of its orbital distance. The brighter the secondary peak, the greater is the mass of the planet. Therefore, by carefully monitoring the timing and intensity of these peaks, it is possible to discover and probe very distant planetary systems.

Doppler spectroscopy can be used to detect planets up to about 100 light years away. The transit technique can detect planets several 100 light years away. In comparison, gravitational microlensing can be used to detect planets thousands of light years away. Indeed, the first planet detected with this technique, designated OGLE–2005-BLG-390LB, is a cold, terrestrial world orbiting a small, $0.22M_{\odot}$ M-dwarf star located near the galactic center, an astounding ~22,000 light years away (!) (Beaulieu et al. 2006; Planetary Society 2022). Since this initial detection in 2006, over 80 exoplanets have been detected using gravitational microlensing. Similar to the transit technique, searches for planets using microlensing can be conducted toward thousands of stars simultaneously, which is good, since the chances of such a fortuitous alignment of stars and observer is quite small. This leads to the biggest downer about this technique, given the relative motion of objects in space, the chance alignment that led to the detection of a planet such as OGLE–2005-BLG-390LB will never happen again, making follow-up observations of these distant worlds virtually impossible.

As discussed in Chapter 7, during the formation of planetary systems, gravitational interactions between protoplanets within a protoplanetary disk can lead to the ejection of planetesimals into cometary Opik- Oort clouds or even interstellar space. One such outcast, a Mars-sized body located many thousands of light years away, was recently discovered via a short timescale (41.5-minute) microlensing event (Mroz et al. 2020). Even more recently, serendipitous space-based observations toward the X-Ray binary source (M51-ULS-1A) in the Whirlpool Galaxy (M51) were indicative of the transit of a Saturn-sized planet (Stefano et al. 2021). These observations further underscore the ubiquity of exoplanets and, with them, the possibility of life elsewhere.

8.3.6 Planets in Habitable Zones

Since the time of Aristarchus (~270 BC), humans (a few at least!) believed they lived on a spherical planet orbiting the Sun and that the stars they saw at night were other suns that harbored planets. Those who held this notion were often held in contempt by their colleagues or even put to death. It was not until the telescopic observations of Galileo and the theoretical work of Kepler and Newton in the 1600s that the tide turned and the heliocentric model of our solar system became generally accepted. Although the belief that planets orbit other stars has been long held, the discovery of the first exoplanet orbiting a main sequence star did not occur until 1994. Since then, due to the use of the transit planet–finding technique on the KEPLER and TESS space telescopes, together with ground-based Doppler spectroscopy, we now have tentative detections of over 4,500 exoplanets. These detections are plotted in Figure 8.8 as a function of parent star mass versus orbital radii. Approximate boundaries of the associated habitable zones are overplotted (Schulze-Makuch et al. 2020). The transit surveys are only sensitive to planets out to a few 100 light years from Earth toward a limited region of the sky. If the survey results are extrapolated to the rest of the Milky Way, the total number of exoplanets is estimated to be over 100 billion. Of these 300 million are estimated to be within the habitable zone of their stars. In some instances, the transit and Doppler spectroscopy techniques can be used together to estimate both the size and mass of a planet. From these a planet's density can be determined. From knowing a planet's density we can tell whether it is a terrestrial planet (~ 5 g/cm^3) or a gas giant (~1 g/cm^3). Based on such analyses, Bryson et al. (2021) estimate

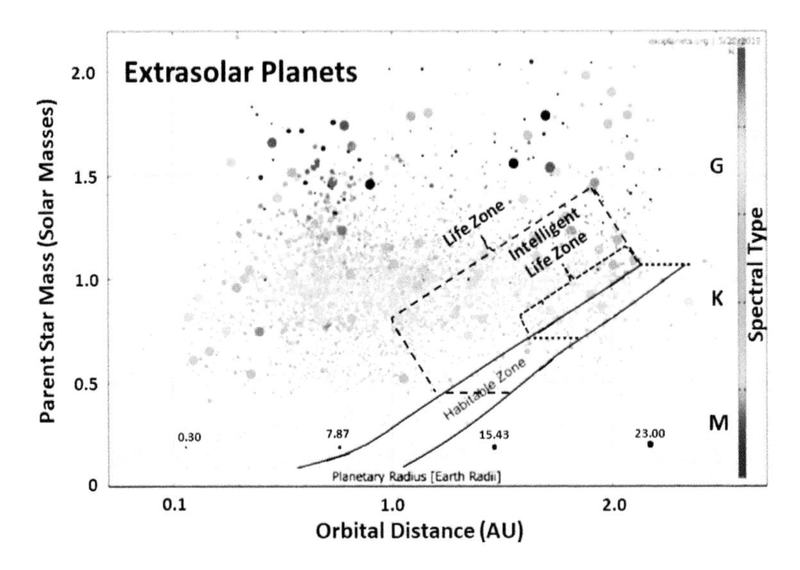

FIGURE 8.8 Extrasolar planets (or exoplanets) and Habitable Zone. Approximately 4,500 exoplanets (confirmed and candidates) are plotted as a function of parent star mass and orbital distance. Colors indicate the parent star spectral type. (Here, colder stars are blue and hotter stars are red.) Symbol diameters are indicative of planet radii. A habitable zone as defined by Kopparapu et al. (2013) is marked with solid black lines. Regions of the habitable zone reflecting the stellar requirements for the evolutionary time scales of life on Earth are indicated with dashed lines. There are ~100 Earth-like planets that fall within the "Intelligent Life Zone" in the plot above, which is ~2% of the total. The actual number of exoplanets in the Milky Way is estimated to be over 100 billion. If the region of outer space sampled to produce the above plot is typical of the galaxy as a whole, then we can estimate the number of planets in the Milky Way with conditions suitable for the potential evolution of intelligent life: >100 billion × 0.02 > 2 billion. Figure adapted from Schulze-Makuch, Heller, and Guinan (2020).

terrestrial planets with radii between 0.37 and 0.6 that of Earth will be found in the habitable zones of K and G stars more than 40% of the time. In the Drake Equation (Chapter 1) this percentage corresponds to a value of $n_e > 0.4$. Given that approximately 12% of all stars are "good stars" (i.e., solitary stars falling into either spectral class G or K; $f_g \approx 0.12$ in the Drake Equation), this value of n_e leads to an estimate for the total number of Earth-like planets in our galaxy of *>5 billion!*

8.4 BIOSIGNATURES

A biosignature is any feature on a planet that can serve as evidence for past or present life. Life had a profound effect on the composition of Earth's atmosphere. Photosynthesis associated with the metabolism of primitive lifeforms changed it from a nitrogen (N_2), carbon dioxide (CO_2), and water (H_2O) atmosphere to one in which molecular oxygen (O_2) is the second most abundant constituent (21% of total). Other molecules that are byproducts of life include methane (CH_4), dimethyl sulfide $(CH_3)_2S$, and chloromethane (CH_3Cl) (Seager 2014; Seager, Bains, and Petkowski 2016). Seeing these in the atmosphere of an exoplanet would provide indirect evidence for life.

The nearest known Earth-like planet that could support intelligent life is Kepler-186f, located ~500 light years away. This seems quite far (and it is, 3,000 *trillion* miles), but in terms of its proximity to us in the Milky Way, it is a next-door neighbor. Even travelling in a starship at warp one (i.e., the speed of light in *Star Trek*), it would take 500 years to get there and once there, the same amount of time to radio your findings back to Earth. So, how can the atmosphere of a planet be detected and probed over interstellar, or, even, interplanetary distances? This was first accomplished on May 26, 1761, by the Russian scientist and poet Mikhail Lomonosov (see Figure 8.9). He did so while observing the transit of Venus across the Sun with a refracting telescope at the St. Petersburg Observatory. Figure 8.10 illustrates the situation. As Venus encroached on the Sun, Lomonosov expected to see the sharp front edge of Venus' circular silhouette. What he actually saw instead is described in this excerpt from the discovery paper (translated by Shiltsev 2014):

FIGURE 8.9 Mikhaylo (Mikhail) Lomonosov (1711–1765). In 1761 he used the transit technique to discover the atmosphere of Venus. Image: USSR stamp. ca 1961: Shutterstock.

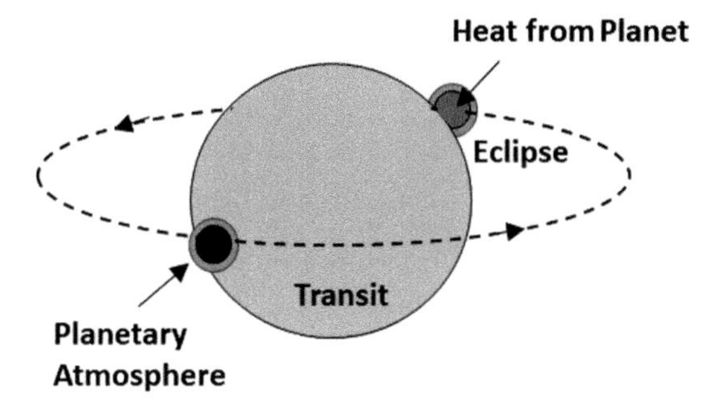

FIGURE 8.10 Transits and eclipses. During transits the planet size can be estimated and its atmosphere measured using light sifting through it from the star. Later as the planet is about to pass behind the star, heat from the surface of the planet can be used to probe the planet's atmosphere in the mid/far infrared.

> "After a few seconds, [Lomonosov] took a glance through the tube and saw that in the place where the Sun's edge had previously appeared somewhat blurred, there was indeed a black pock or segment, which was very small, but no doubt due to the encroaching Venus. Then [he] watched attentively for the entry of the other (trailing) edge of Venus, which seemed to have not yet arrived, and a small segment remained beyond the Sun. However, then suddenly there appeared between the entering trailing edge of Venus and the solar edge, a hair-thin bright radiance separating them, so that the time from the first to the second was no more than one 95 second … From these observations, Mr. Councilor Lomonosov concludes that the planet Venus is surrounded by a significant air atmosphere [that refracts light] similar to (if not even greater than) that which surrounds our terrestrial globe."

Lomonosov was right. Venus does have an atmosphere even denser than our own. He then goes on to speculate about the possibility of life on Venus.

> "After reading here about the great atmosphere around the aforementioned planet, one can say: we can then presume that because of its vapor updrafts, clouds gather, rains fall, flowing streams gather into the rivers, and the rivers flow into the seas, different types of plants grow everywhere on which animals feed."

While life may have been spawned on Venus early in its history, the same thick atmosphere responsible for the strong refraction of sunlight observed by Lomonosov now makes life on the Venusian surface problematic at best (see Chapter 6). Today we can use the transit technique pioneered by Lomonosov over 250 years ago to not only search for planets but also to look for biosignatures in their atmospheres (Lorenz 2019).

Lomonosov observed sunlight through the Venusian atmosphere, "a hair-thin brightness radiance". If we pass such light through a spectrograph, we will see the absorption spectrum of the background star (in our case the Sun). Superimposed on the stellar spectrum will be a second, faint absorption line spectrum due to light being preferentially absorbed by molecules within the intervening planetary atmosphere (see Figure 8.10). Looking back at Figure 8.6 we can see that when the planet goes behind the star, the light from the planet is completely blocked and we see only the stellar spectrum. We can then subtract the pure stellar spectrum from the transit observation to unmask the absorption line spectrum of the planet's atmosphere. We can then compare the planetary absorption line spectrum to laboratory spectra to identify the molecules within the planet's atmosphere … this is truly amazing. We can potentially probe planetary atmospheres using a combination of space and ground-based telescopes with little more than a laptop computer and a decent internet connection. Such exoplanet surveys are now underway.

Compared to the underlying stellar spectrum, the planetary atmospheric spectrum is extremely weak. To subtract the two without adding noise that masks biosignatures in a planet's atmospheres requires detector stabilities measured in parts per million. Such detectors will soon be available for the next-generation space telescopes. Due to their quantum

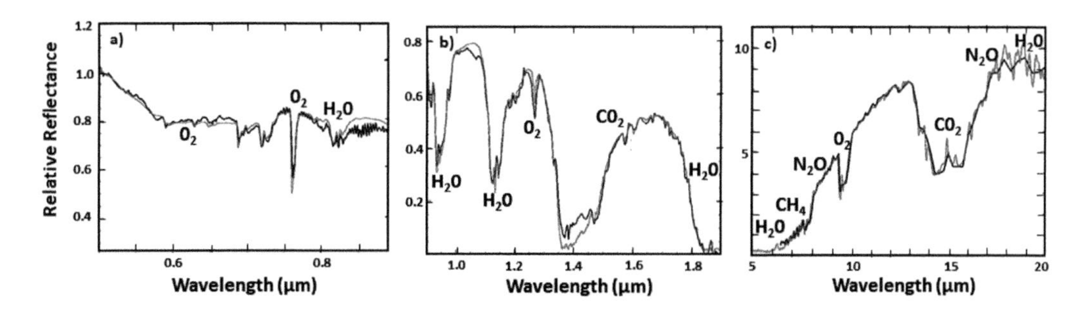

FIGURE 8.11 Earth as an exoplanet. (a) visible, (b) near-infrared reflectivity, (c) and infrared emission spectra of Earth. Atmospheric model results are in blue. Black lines are from Earthshine (reflected light from the Moon) measurements and red from spacecraft measurements. Figure after Kaltenegger et al. 2009.

mechanical structure many biogenic molecules have absorption line features at near- and mid-infrared wavelengths, between ~2 and 30 microns. The spectrum of Earth (our gold standard) over this range is shown in Figure 8.11. A number of biogenic molecules are seen. If you were an alien looking at this spectrum, what would suggest the presence of life and habitability would be, in addition to H_2O and CO_2, the presence of O_2 and O_3 in combination with CH_4 or N_2O (Kaltenegger 2017). The best atmospheric biosignature would be to detect a biogenic gas out of chemical equilibrium and many orders of magnitude higher than expected from nonbiological processes (e.g., of photochemical or geological origins).

Let's consider the case of O_2. Molecular oxygen is a highly reactive–corrosive gas that reacts readily with surface minerals, as well as volcanic gases (e.g., H_2 and H_2S). If there was a mass extinction event on Earth (e.g., collision with a sufficiently large left-over planetesimal; comet, asteroid, or meteor) and the biosphere collapsed, there would be no replenishment of O_2 and it would dissipate within a few million years (Kaltenegger et al. 2009). An alien observer might then incorrectly conclude the Earth had always been devoid of life.

As we discussed in Chapter 5, life has dramatically transformed the Earth's atmosphere since its formation 4.7 billion years ago. Initially the atmospheric composition was dominated by geological processes (volcanic outgassing of N_2, H_2O, and CO_2). Overtime a significant fraction of the CO_2 was washed out of the atmosphere by rain and subsequently accumulated in surface rocks and the ocean. With the advent of organisms capable of photosynthesis, the oxygen level slowly began to rise. It took billions of years of atmospheric evolution to yield the observed spectrum in 8.11. Figure 8.12 shows snapshots of how the Earth's atmosphere would appear to aliens in the past. Such evolution should be taken into account when considering the habitability of a world today or in the past.

There are at least 1,004 main sequence stars within 326 light years (100 pc) of us from which neighboring civilizations (if they exist) in orbit about them could have observed transit spectra of the Earth against the Sun like those of Figures 8.11 and 8.12. Of these 77% are M-type, 12% K-type, 6% G-type, 4% F-type stars, and 1% A-type stars. The closest of these is 28 light years away (Kaltenegger and Pepper 2020). One cannot help but wonder if such observations have motivated investigation of the Earth by others.

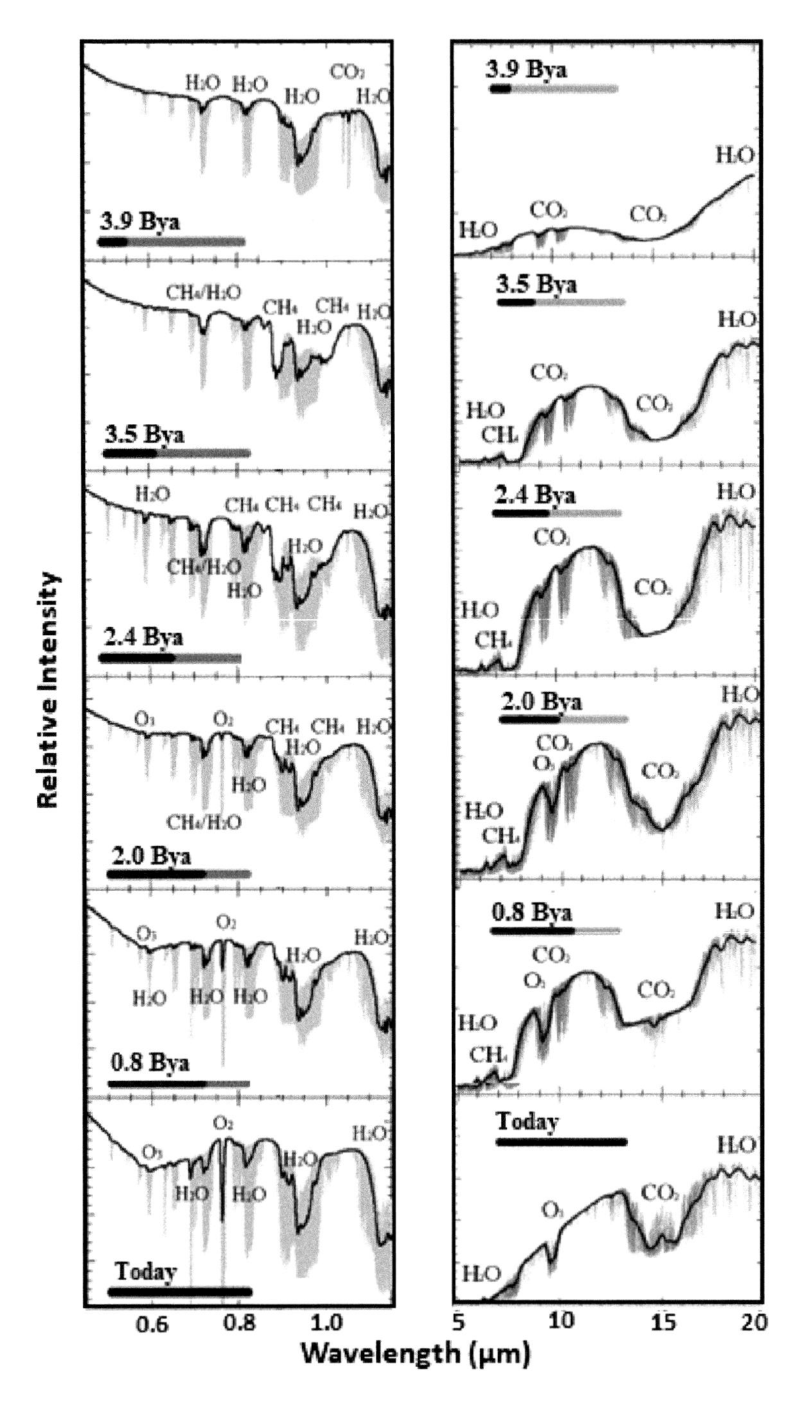

FIGURE 8.12 Spectroscopic evolution of the Earth's atmosphere. The composition of the Earth's atmosphere has evolved from being dominated by geological process 3.9 billion years ago (top) to today's atmosphere (bottom) where the influence of life can be clearly seen. After Figures 9b and 10b (with clouds) of Kaltenegger 2007, with permission.

8.5 M-STAR ADVANTAGE

The signal for detecting a transiting planet is directly proportional to the projected area of the planet, A_p relative to that its star, A_S. Since terrestrial planets are relatively small, they are easier to detect against the background of a smaller star, such as M-dwarfs. Also, results from the Kepler mission suggest that the number of small planets per star increases for smaller, cooler stars (Dressing and Charbonneau 2015). As we learned in Chapter 3, there are many more low-mass stars than high-mass stars. Due to their low luminosity, the habitable zones for M-dwarfs are located close-in to the stars (see Figure 8.3). From Kepler's Laws this means their orbital periods are relatively quick, 4–20 days, increasing the number of transit opportunities. All these factors make M-dwarf stars a natural subject of exoplanet studies.

One such M-dwarf is TRAPPIST-1, located approximately 40 light years away (de Wit et al. 2018). In 2017 *seven* terrestrial planets were found orbiting Trappist-1 using the transit technique with the Hubble Space Telescope (Gillon et al. 2017). Three of the seven worlds are located within the star's habitable zone (see Figure 8.13). Of these, TRAPPIST-1e is most similar to Earth in mass ($0.772\,M_\oplus$), density ($0.98\,\rho_\oplus$), size ($0.910\,R_\oplus$), and temperature(–27.1°C; –16.7°F). The temperature estimate comes from its distance from the star. Greenhouse gases may raise its temperature above freezing, as they do over much of the Earth. Subsequent TTV studies of the TRAPPIST system are consistent with planets c and e having large rocky interiors, while b, d, f, g, and h require envelopes of volatiles, which can take the form of thick atmospheres, oceans, or ice (Grimm et al. 2018). Early observations suggest TRAPPIST-1e may have a compact atmosphere and perhaps an ocean. Figure 8.14 is a computer simulation of the anticipated transmission spectrum taken during transit of the TRAPPIST-1e atmosphere assuming it is cloud-free and has a CO_2 partial pressure similar to that of Earth. Model runs were made assuming it is wet and alive, wet and dead, and dry and dead. The presence of O_2, CH_4, and NO_2 is particularly

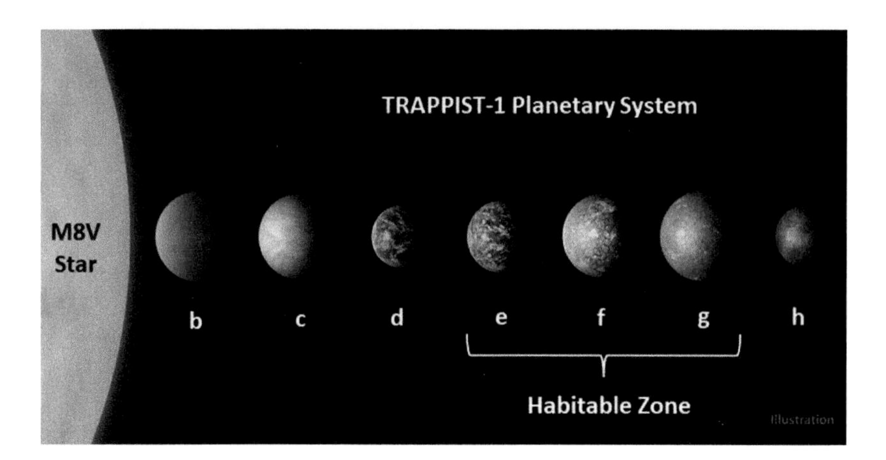

FIGURE 8.13 TRAPPIST-1 Planetary System. The system contains seven terrestrial worlds, three of which are within the M8V star's habitable zone. TRAPPIST-1e is the most Earth-like. Image Credit: NASA.

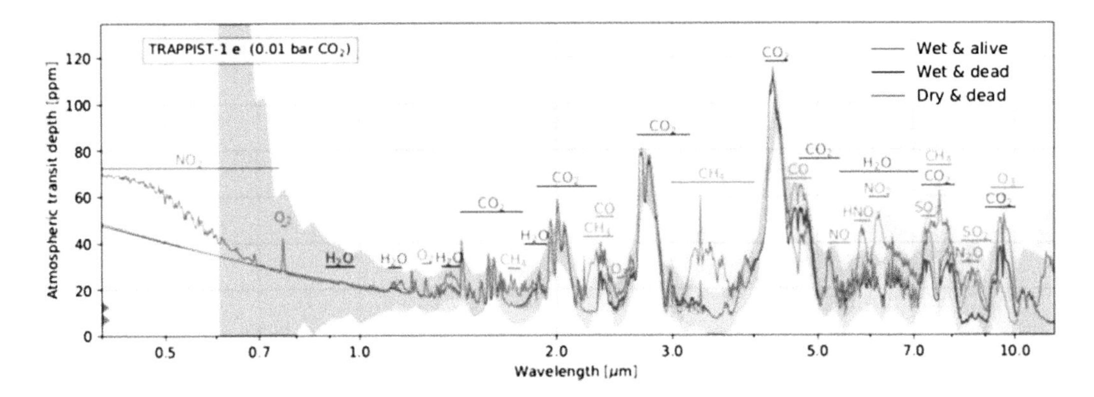

FIGURE 8.14 Simulated spectrum of TRAPPIST-1e assuming a partial pressure for CO_2 of 0.01 bar. Computer model runs were made assuming the planet is wet and alive (green line), wet and dead (blue line), and dry and dead (red line). The shaded areas indicate the 1 sigma noise level expected from co-adding 30 transit observations made with JWST. Even with this powerful telescope, it will be difficult to discern between the different cases. In order to reach the necessary sensitivity and to robustly estimate the number of planets that harbor life, even more powerful telescopes will be required. Adapted from Wunderlich et al. 2020.

FIGURE 8.15 James Webb Space Telescope (JWST). Launched in December 2021, the cooled, 6.5 meter diameter mirror of JWST, together with its suite of advanced instrumentation, will provide new insights into the origin and evolution of planetary systems, stars, galaxies, and the Universe itself. Image Credit: NASA.

important for determining if life, as we know it (as indicated by the green line), may exist there (Wunderlich et al. 2020). In order to tell the difference between the various cases, multiple observations will be required at near and mid-infrared wavelengths using the most powerful space telescopes. Currently, this is the James Webb Space Telescope (JWST; see Figure 8.15).

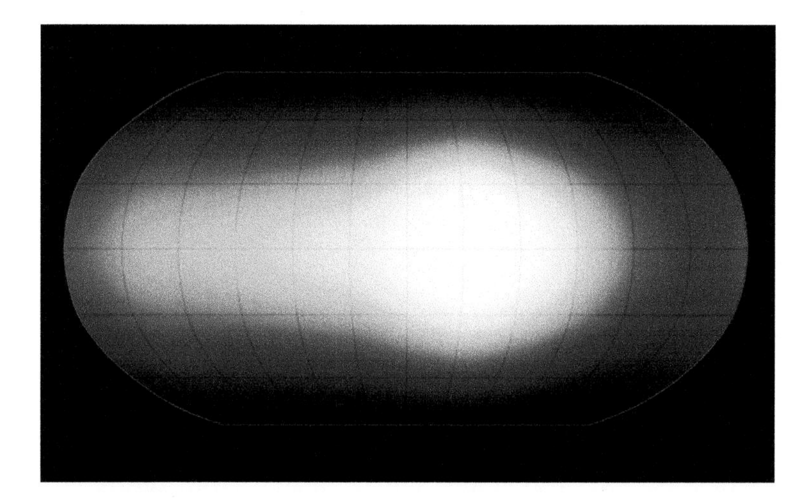

FIGURE 8.16 Global temperature map of the "hot-Jupiter" HD189773b. The map was derived by monitoring the infrared radiation from the planet over half an orbital period as the dayside of the planet rotated into view. The off-center peak in temperature suggests the presence of atmospheric winds. Image Credit: NASA/Knutson et al. 2007.

In addition, by observing the infrared light emitted from the dayside of the planet before it passes behind the star (i.e., goes into an eclipse), it is possible to create rudimentary temperature maps of the planet (see Figure 8.16). If, as one sees here, the peak in the temperature map is not exactly facing the star, then this could suggest the presence of strong, atmospheric circulation patterns, which when combined with temperature and pressure data, can be used to study the weather on distant worlds (Knutson et al. 2007; Ofir 2016).

8.6 THE CASE OF PROXIMA B

Our nearest stellar neighbor is Proxima Centauri, a small red dwarf, M5.5 star with a luminosity of only 0.0017 of the Sun. At a distance of only 4.2465 light years, it is within the interstellar "spitting distance" of the Earth (Turbet et al. 2016). Using Doppler spectroscopy, three planets have been detected orbiting Proxima Centauri; they are Proxima b, Proxima, c, and Proxima d. Analysis of the data reveals that Proxima b orbits Proxima Centauri every 11.2 days at a distance of 0.05 AU, which places it within the star's habitable zone (Anglada-Escudé 2016). Its mass comes in at $1.17M_\oplus$. (You would weigh 17% more there.) In orbiting so close to the star, there is a chance Proxima b is tidally locked, like the Moon is to the Earth, with one side permanently in daylight and the other in night. Surface habitability would require the planet to have both an atmosphere and a mild greenhouse effect for liquid water to exist. An atmosphere would also help redistribute thermal energy around the planet and thereby moderate temperature extremes. Another issue for habitability is that Proxima Centauri has frequent solar flares, i.e., coronal mass ejection (CME) events that may have overtime eroded away Proxima b's atmosphere (Khodachenko et al. 2007). If a planet has a strong enough magnetic field, the field could deflect the oncoming deluge of charged particles, thereby preventing or slowing atmospheric erosion. As we learned in Chapter 5, Mars does not have a significant global magnetic field and, as a result, may have lost its atmosphere early on due to erosion by the solar wind, leading to it being a frozen world. If the core region of Proxima b is molten, then since the planet is tidally locked, one rotation per orbital period could cause the core to rotate within the planet. Such a rotation may be sufficient to produce a protective magnetosphere. Being tidally locked and suffering atmospheric erosion from solar flares may be a hindrance to habitability common to terrestrial planets orbiting M-dwarf stars.

As fate would have it, the orientation of the orbital plane of Proxima b is tilted sufficiently with respect to the Earth that the planet does not transit the star (Gilbert et al. 2021). Therefore, we cannot use the transit technique to perform spectroscopic analysis of its atmosphere as we can for TRAPPIST-1e. However, its angular separation from Proxima is sufficient that it could, in principle, be directly observed in the infrared (where the contrast between planet and star is greatest) with future large telescopes, either from the ground using adaptive optics (e.g., the 40 m European Extremely Large Telescope) or from space.

Given its close proximity, another exciting possibility is to actually go there. This could be accomplished using spacecraft with nuclear propulsion systems (e.g., Project Orion and Project Daedalus; see Long 2011). Such interstellar spacecraft could be available by the end of this century. In the meantime, plans are underway to develop light sail interstellar probes (see Figure 8.17) that would use a ground-based, multi-kilometer phased array of lasers to provide the photonic wind needed to drive the spacecraft to their destination. The project's name is Breakthrough Starshot (Parks 2021; Lubin 2016). Using this technical approach, it would be possible to accelerate small, low mass (<1 gm) "wafersats" to speeds exceeding 0.25c (i.e., 25% the speed of light) and reach the Proxima system in ~15 years.

FIGURE 8.17 Solar sail concept. The propulsion system for our first interstellar probe to a nearby world (Proxima b) could use a light sail driven by photon pressure produced by powerful lasers. Image Credit: Shutterstock.

8.7 THE CASE OF MARS

In 1976, three life detection experiments were performed on Martian soil samples by the Viking 1 and 2 landers (see Figures 8.18 and 8.19). One of them, the Labeled-Release experiment was designed to detect metabolism associated with microbial life. It gave positive results. However, the other two experiments, which searched for organic molecules, showed negative. Whether life was detected by the Viking landers remains a subject of debate (Bianciardi et al. 2012). Since Viking there have been a number of unmanned missions to Mars; including orbiters, landers, and rovers. One of the most intriguing discoveries was made by the Mars Reconnaissance Orbiter (MRO), which observed narrow, 100-meter-long dark steaks running down the sides of hills on Mars (see Figure 5.20). The streaks appear darker during warm seasons, when temperatures can reach −10°F (−23°C)

FIGURE 8.18 Viking Orbiter "Mothership". Two Viking motherships entered Martian orbit in the summer of 1976. Each mothership consisted of an orbiter and a lander. The landers were housed in a protective aeroshell (that looked a lot like a flying saucer) for entry into the Martian atmosphere. At an altitude of 3.7 miles (600 km) a parachute was released that slowed decent and the saucer opened up like a clamshell exposing the lander. The lander then extended its legs and descent engines brought it to a soft landing on the surface. Because of the time lag in radio transmissions due to distance, the landing operation was performed autonomous by onboard computers. Image Credit: Don Davis; Public Domain.

FIGURE 8.19 Viking 2 lander image. This image was taken on the surface of Mars in 1976. The Viking landers carries three biological experiments designed to detect the presence of microbial life. The results were inconclusive. Image: NASA.

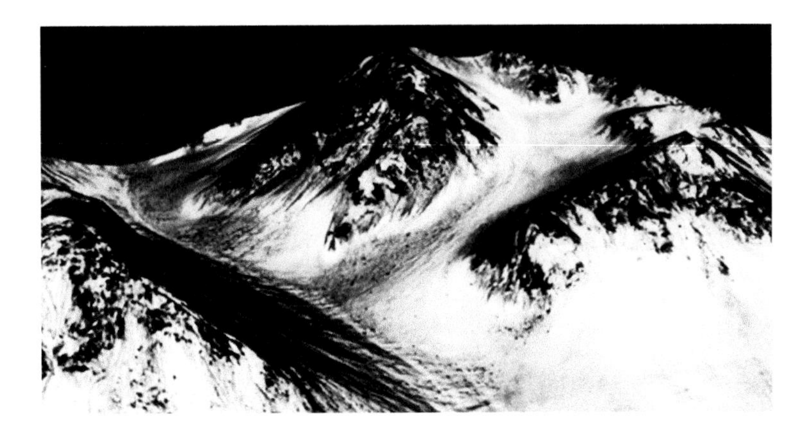

FIGURE 8.20 Dark streaks on the side of Hale Crater. One interpretation is that these seasonal narrow, 100 m long dark streaks may be the result of liquid water running downhill during the Martian summer. Alternatively, they could be the results of granular flows, where grains of sand and dust slip downhill to make dark streaks, rather than the ground being darkened by seeping water. Image: NASA/JPL/University of Arizona.

and fade during cold seasons. Hundreds of individual streaks may be observed over a local slope and thousands in individual images captured by the High-Resolution Imaging Science Experiment on MRO (HiRISE; McEwen et al. 2007). Spectroscopic analysis of the streaks (referred to as recurring slope linear (RSL)) made from MRO indicate the presence of hydrated salts in their vicinity. The salts could have the effect of lowering the freezing point of water, allowing it to exist in a liquid state during the relatively cold Martian summers (Chevrier and Rivera-Valentin 2012). The streaks could then be the result of brine water running downhill. Alternative explanations for the streaks are that they are granular flows, where grains of sand and dust slip downhill to make dark streaks, rather than the ground being darkened by seeping water (Schmidt et al. 2017: McEwen et al. 2021). On-site investigation of RSLs at multiple locations by unmanned rovers or astronauts will likely be required to definitively solve the mystery of their origin. However, water ice is known to be present just below the surface of the Phoenix lander site (Smith et al. 2009), as well as several other mid-latitude sites observed from orbit by HiRISE or through Gamma Ray Spectrometer (GRS) data (Bishop et al. 2021; Piqueux et al. 2019). On Earth, where there is liquid water, there is life. This is even the case in the McMurdo Dry Valleys of Antarctica, where endolithic bacteria (i.e., bacteria living within rock) can be found that draw nutrients from summer melt water from nearby glaciers (Barrett et al. 2007). A much more extensive study of Martian soil samples from multiple locations and depths is required to answer the age-old question of whether life exists or has existed on Mars. The presence of water on Mars gives us hope we may yet find evidence of fellow creatures (however diminutive) living on or in other worlds within our own solar system.

At the time of this writing the car-sized NASA rover (Perseverance) and its sidekick reconnaissance helicopter (Ingenuity) are exploring the 45 km wide Jerzero crater. The crater is suspected of once hosting a long-lived lake. Perseverance has four objectives: (1) looking for habitability, (2) seeking biosignatures, (3) caching samples for a future sample-return mission, and (4) testing oxygen production from the Martian atmosphere (NASA 2022). The rover's prime target is a fan-shaped pile of sediments that washed into the crater ~3.5 billion years ago (see Figure 8.21; Drake 2022; Witze 2022). The samples are expected to be returned to Earth for analysis in the early 2030s. Perhaps then the question of the existence of life beyond Earth will be answered.

8.8 SUMMARY

Humans have contemplated the possibility of planets orbiting other stars for well over 2,000 years. Isaac Newton first computed the habitable zone for planets orbiting the Sun in the seventeenth century. The atmosphere of the first extraterrestrial rocky planet (Venus) was serendipitously discovered by Mikhail Lomonosov about 100 years later. For the next two centuries, methods for detecting extrasolar planets were proposed and attempted. These included direct imaging, astrometry, and Doppler spectroscopy. But it was not until the 1990s that technologies matured to a sufficient level to detect the first Jupiter-sized planets. These Jovian worlds were found to be orbiting surprisingly close to their stars, making them easier to detect than expected. This led to a revision in planet formation theory and a "gold rush" in exoplanet research, subsequently leading to the discovery of thousands of exoplanets with the Kepler Space Observatory using the transit technique first employed by Lomonosov over two centuries earlier. Planets ranging from small Mars-size bodies to super-Jupiters have been found orbiting at varying distances from their stars. Statistical analyses have revealed that the fraction of "good stars" with Earth-mass planets orbiting in habitable zones is ~0.4. This then leads to an estimate for the number of

FIGURE 8.21 Perseverance landing site (dashed circle). The fan-shaped delta was formed ~3.5 billion years ago by water breaking through the rim of the Jerzero crater. The false color image combines data from two instruments on the Mars Reconnaissance Orbiter. Image Credit; NASA/JPL/JHUAPL/MSSS/Brown University.

potentially habitable planets in our galaxy of more than 5 billion. Using a new generation of telescopes (e.g., JWST) and detection systems, it is now becoming possible to probe the chemical make-up of relatively nearby exoplanet atmospheres for molecular biosignatures (Walker et al. 2018). Once a statistically significant number of exoplanet atmospheres have been studied, we will be able to robustly estimate the fraction of habitable planets in the Milky Way on which life has evolved. Profound insights into this question will also be gained through our continued search for life in the solar system. A positive result would be definitive proof that the occurrence of life is not limited to Earth.

In the previous chapter, we learned that once life originates on a planet there is an evolutionary pressure toward greater complexity. On Earth, it took ~4 billion years before humans appeared. Did evolution on other planets lead life down similar biological, cultural, and technological paths, or in completely different directions? The drive to answers these questions is ingrained in our DNA and collective conscientiousness. This innate desire will continue to motivate our investigation of life in the universe for generations to come.

REVIEW QUESTIONS

1. What are the different techniques used for finding planets around other stars (i.e., exoplanets)?

2. Which technique for finding planets has been the most successful? Why?

3. Why are massive planets close to a star easier to detect?
4. Why is direct imaging of planetary systems so hard?
5. Which technique can be used to detect planets halfway across the galaxy?
6. Why is atmospheric pressure important to the likelihood of a planet having liquid water on its surface?
7. What role does a planet's atmosphere play in determining its surface temperature?
8. Why is planetary rotation important to estimating the surface temperature of a planet?
9. About how many extrasolar planets have been detected so far?
10. What is the habitable zone (or ecosphere) of a star?
11. Who first estimated the habitable zone of our solar system?
12. What is the surface temperature and atmospheric pressure on Venus, Earth, and Mars?
13. What is currently the habitable zone of our Sun?
14. How do the boundaries of a habitable zone change over time?
15. Which spectral type star has the widest habitable zone?
16. Which spectral type star has the smallest habitable zone?
17. What range of spectral type stars are more likely to host planets with life? How many of these stars are estimated to be in the Milky Way?
18. What range of spectral type stars are more likely to host planets with intelligent life? How many of these stars are estimated to be in the Milky Way?
19. Who first observed an atmosphere around Venus?
20. From this first observation what was concluded about the nature of the Venusian atmosphere?
21. What is a planetary biosignature?
22. What spectral features in the atmosphere of a planet may be indicative of life?
23. What techniques are available to search for biosignatures in other planetary systems?
24. What is the nearest known exoplanet that may be capable of supporting life?
25. How long would it take to get there in a starship travelling at 10% the speed of light?
26. How many planets in the Milky Way may be capable of supporting life?

27. What is the nearest known exoplanet that may be capable of harboring indigenous intelligent life?
28. How many planets in the Milky Way may be capable of harboring indigenous intelligent life?
29. How can surface temperature maps of an exoplanet be made?
30. Why would finding indigenous life on another planet in our solar system be so important?
31. What evidence is there that life may exist or has existed on Mars?

REFERENCES

Anglada-Escude, G., P. Amado, J. Barnes, Z. M. Berdiñas, R. P. Butler, G. A. L. Coleman, I. de la Cueva, S. Dreizler, M. Endl, B. Giesers, S. V. Jeffers, J. S. Jenkins, H. R. A. Jones, M. Kiraga, M. Kürster, M. J. López-González, C. J. Marvin, N. Morales, J. Morin, R. P. Nelson, J. L. Ortiz, A. Ofir, S. Paardekooper, A. Reiners, E. Rodríguez, C. Rodríguez-López, L. F. Sarmiento, J. P. Strachan, Y. Tsapras, M. Tuomi, and M. Zechmeister. 2016. "A Terrestrial Planet Candidate in a Temperate Orbit around Proxima Centauri." *Nature* 536 (7617): 437.

Barrett, J. E., R. A. Virginia, W. B. Lyons, D. M. McKnight, J. C. Priscu, P. T. Doran, A. G. Fountain, D. H. Wall, and D. L. Moorhead. 2007. "Biogeochemical Stoichiometry of Antarctic Dry Valley Ecosystems." *Journal of Geophysical Research: Biogeosciences* 112: G01010.

Beaulieu, J.-P., D. P. Bennett, P. Fouqué, A. Williams, M. Dominik, U. G. Jørgensen, D. Kubas, A. Cassan, C. Coutures, J. Greenhill, K. Hill, J. Menzies, P. D. Sackett, M. Albrow, S. Brillant, J. A. R. Caldwell, J. J. Calitz, K. H. Cook, E. Corrales, M. Desort, S. Dieters, D. Dominis, J. Donatowicz, M. Hoffman, S. Kane, J.-B. Marquette, R. Martin, P. Meintjes, K. Pollard, K. Sahu, C. Vinter, J. Wambsganss, K. Woller, K. Horne, I. Steele, D. M. Bramich, M. Burgdorf, C. Snodgrass, M. Bode, A. Udalski, M. K. Szymański, M. Kubiak, T. Więckowski, G. Pietrzyński, I. Soszyński, O. Szewczyk, Ł. Wyrzykowski, B. Paczyński, F. Abe, I. A. Bond, T. R. Britton, A. C. Gilmore, J. B. Hearnshaw, Y. Itow, K. Kamiya, P. M. Kilmartin, A. V. Korpela, K. Masuda, Y. Matsubara, M. Motomura, Y. Muraki, S. Nakamura, C. Okada, K. Ohnishi, N. J. Rattenbury, T. Sako, S. Sato, M. Sasaki, T. Sekiguchi, D. J. Sullivan, P. J. Tristram, P. C. M. Yock, and T. Yoshioka. 2006. "Discovery of a Cool Planet of 5.5 Earth Masses through Gravitational Microlensing." *Nature* 439 (7075): 437–40.

Benedict, G., McArthur, B., Forveille, T., and 9 others. 2002. "A Mass for the Extrasolar Planet Gliese 876b Determined from Hubble Space Telescope Fine Guidance Sensor 3 Astrometry and High-Precision Radial Velocities." *Astrophysical Journal Letters* 581 (2): L115–L8.

Bianciardi, G., J. D. Miller, P. A. Straat, and G. V. Levin. 2012. "Complexity Analysis of the Viking Labeled Release Experiments." *IJASS* 13 (1): 14–26.

Bishop, J. L., M. Yesilbas, N. W. Hinman, Z. F. Burton, M., P. A. J. Englert, J. D. Toner, A. S. McEwen, V. C. Gulick, E. K. Geibson, and C. Koeberl. 2021. "Martian Subsurface Cryosalt Expansion & Collapse as Trigger for Landslides." *Science Advances* 7: eabe4459.

Borucki, W. J., D. Koch, G. Basri, N. Batalha, T. Brown, D. Caldwell, J. Caldwell, J. Christensen-Dalsgaard, W. D. Cochran, E. DeVore, E. W. Dunham, A. K. Dupree, T. N. Gautier, J. C. Geary, R. Gilliland, A. Gould, S. B. Howell, J. M. Jenkins, Y. Kondo, D. W. Latham, G. W. Marcy, S. Meibom, H. Kjeldsen, J. J. Lissauer, D. G. Monet, D. Morrison, D. Sasselov, J. Tarter, A. Boss, D. Brownlee, T. Owen, D. Buzasi, D. Charbonneau, L. Doyle, J. Fortney, E. B. Ford, M. J. Holman, S. Seager, J. H. Steffen, W. F. Welsh, J. Rowe, H. Anderson, L. Buchhave, D. Ciardi, L. Walkowicz, W. Sherry, E. Horch, H. Isaacson, M. E. Everett, D. Fischer, G. Torres, J.

A. Johnson, M. Endl, P. MacQueen, S. T. Bryson, J. Dotson, M. Haas, J. Kolodziejczak, J. Van Cleve, H. Chandrasekaran, J. D. Twicken, E. V. Quintana, B. D. Clarke, C. Allen, J. Li, H. Wu, P. Tenenbaum, E. Verner, F. Bruhweiler, J. Barnes, and A. Prsa. 2010. "Kepler Planet-Detection Mission: Introduction and First Results." *Science* 327 (5968): 977–80.

Bryson, S., M., Kunimoto, R., Kopparapu, and 79 others. 2021. "The Occurrence of Rocky Habitable-zone Planets around Solar-like Stars from Kepler Data." *AJ* 161: 36.

Charbonneau, D., T. M. Brown, D. W. Latham, and M. Mayor. 2000. "Detection of Planetary Transits Across a Sun-Like Star." *Astrophysical Journal* 529 (1): L45–L8.

Chauvin, G., A. M. Lagrange, C. Dumas, B. Zuckerman, D. Mouillet, I. Song, J. L. Beuzit, and P. Lowrance. 2004. "A Giant Planet Candidate near a Young Brown Dwarf. Direct VLT/NACO Observations Using IR Wavefront Sensing." *Astronomy & Astro-Physics* 425 (2): L29–L32.

Chevrier, V. F., and E. G. Rivera-Valentin. 2012. "Formation of Recurring Slope Lineae by Liquid Brines on Present-Day Mars." *Geophysical Research Letters* 39 (21): L21202.

de Wit, J., H. R. Wakeford, N. K. Lewis, L. Delrez, M. Gillon, F. Selsis, J. Leconte, B. Demory, E. Bolmont, V. Bourrier, A. J. Burgasser, S. Grimm, E. Jehin, S. M. Lederer, J. E. Owen, V. Stamenković, and A. H. M. J. Triaud. 2018. "Atmospheric Reconnaissance of the Habitable-Zone Earth-Sized Planets Orbiting TRAPPIST-1." *Nature Astronomy* 2 (3): 214–9.

Drake, N. 2022. "What We Learned from the Perseverance Rover's First Year on Mars." *Scientific American*, February 17.

Dressing, C. and D., Charbonneau. 2015. "The Occurrence of Potentially Habitable Planets Orbiting M Dwarfs Estimated from the Full Kepler Dataset and an Empirical Measurement of the Detection Sensitivity." *ApJ*, 897:45D.

Evans, N. 1996. *Extraterrestrial Life*. Edina, Minnesota: Burgess Publishing.

Gilbert, E., T., Barclay, E., Kruse, E., Quintana, and L., Walkowicz. 2021. "No Transits of Proxima Centauri Planets in High-Cadence TESS Data." *Frontiers in Astronomy and Space Sciences* 8(769371).

Gillon, M., A. Triaud, B. Demory, E. Jehin, E. Agol, K. M. Deck, S. M. Lederer, J. de Wit, A. Burdanov, J. G. Ingalls, E. Bolmont, J. Leconte, S. N. Raymond, F. Selsis, M. Turbet, K. Barkaoui, A. Burgasser, M. R. Burleigh, S. J. Carey, A. Chaushev, C. M. Copperwheat, L. Delrez, C. S. Fernandes, D. L. Holdsworth, E. J. Kotze, V. Van Grootel, Y. Almleaky, Z. Benkhaldoun, P. Magain, and D. Queloz. 2017. "Seven Temperate Terrestrial Planets around the Nearby Ultracool Dwarf Star TRAPPIST-1." *Nature* 542 (7642): 456–60.

Grimm, S., B-O, Demory, M., Gilon, and 23 others. 2018. "The nature of the TRAPPIST-1 exoplanets." *A&A* 613A: 68G.

Huang, S.-S. 1959. "Occurrence of Life in the Universe." *American Scientist* 47: 397–402.

Jacob, W. S. 1855. "On Certain Anomalies Presented by the Binary Star 70 Ophiuchi." *Monthly Notices of the Royal Astronomical Society* 15 (9): 228–30.

Kaltenegger, L. 2017. "How to Characterize Habitable Worlds and Signs of Life." *Annual Review of Astronomy & Astrophysics* 5 (1): 433–85.

Kaltenegger, L., and J. Pepper. 2020. "Which Stars Can See Earth as an Exoplanet." *Monthly Notices of the Royal Astronomical Society* 499 (1): L111–L5.

Kaltenegger, L., W. Traub, and K. Jucks. 2007. "'Spectral Evolution of an Earth-like Planet'." *The Astrophysical Journal* 658: 598.

Khodachenko, M. L., H. Lammer, J. Grießmeier, M. Leitner, F. Selsis, C. Eiroa, A. Hanslmeier, H. K. Biernat, H. K. Biernat, C. J. Farrugia, and H. O. Rucker. 2007. "Coronal Mass Ejection (CME) Activity of Low Mass M Stars as an Important Factor for the Habitability of Terrestrial Exoplanets. I. CME Impact on Expected Magnetospheres of Earth-Like Exoplanets in Close-In Habitable Zones." *Astrobiology* 7 (1): 167–84.

Knutson, H., D. Charbonneau, L. Allen, J. Fortney, J. Agol, N. Cowan, A. Showman, C. Cooper, and T. Megeath. 2007. "A Map of the Day–Night Contrast of the Extrasolar Planet HD 189733b." *Nature* 447 (7141): 183.

Kopparapu, R. K., R. Ramirez, J. F. Kasting et al. 2013. "'Habitable Zones Around Main-Sequence Stars: New Estimates'." *The Astrophysical Journal* 765: 131.
Lingham, M. 2021. "'A Brief History of the Term 'Habitable Zone' in the 19th Century." *International Journal of Astrobiology* 20 (5): 332–6.
Long, K. F. 2011. *Deep Space Propulsion: A Roadmap to Interstellar Flight*. New York, NY: Springer.
Lorenz, R. 2019. "A Bayesian Approach to Biosignature Detection on Ocean Worlds." *Nature Astronomy* 3 (6): 466.
Lubin, P. 2016. "A Roadmap to Interstellar Flight." *Journal of the British Interplanetary Society* 69: 40.
Mayor, M., and D. Queloz. 1995. "A Jupiter-Mass Companion to a Solar-Type Star." Nature 378 (6555): 355–9.
McEwen, A. S., E. M. Eliason, J. W. Bergstrom, N. T. Bridges, C. J. Hansen, W. Alan Delamere, J. A. Grant, V. C. Gulick, K. E. Herkenhoff, L. Keszthelyi, R. L. Kirk, M. T. Mellon, S. W. Squyres, N. Thomas, and C. M. Weitz. 2007. "Mars Reconnaissance Orbiter's High Resolution Imaging Science Experiment (HiRISE)." *Journal of Geophysical Research* 112: E05S02.
McEwen, A. S., E. I. Schaefer, C. M. Dundas, S. S. Sutton, L. K. Tamppari, and M. Chojnacki. 2021. "Mars: Abundant Recurring Slope Lineae (RSL) Following the Planet-Encircling Dust Event (PEDE) of 2018." *Journal of Geophysical Research: Planets* 126 (4): e2020JE006575.
Mroz, P. et al. 2020. "'A Terrestrial-Mass Rogue Planet Candidate Detected in the Shortest-Timescale Microlensing Event'." *The Astrophysical Journal Letters* 903: L11.
NASA. 2022. https://mars.nasa.gov/mars2020/mission/overview/.
Newton, I. 1687. *Philosophiæ Naturalis Principia Mathematica*. London: Royal Society.
Ofir, A. 2016. "Planetary Transits: How Can One Measure the Mass, Density, Size and Atmospheric Composition of a Planet One Cannot Even See?" *Pale Red Dot*. Accessed March 16, 2016. https://palereddot.org/planetary-transits-how-can-one-measure-the-mass-size-density-and-atmospheric-composition-of-a-planet-one-cannot-even-see/.
Parks, J. 2021. "Breakthrough Starshot: A Voyage to the Stars within Our Lifetimes." *Astronomy*, May. https://astronomy.com/magazine/news/2021/06/breakthrough-starshot-a-voyage-to-the-stars-within-our-lifetimes.
Piqueux, S., J. Buz, C. S. Edwards, J. L. Bandfield, A. Kleinböhl, D. M. Kass, and P. O. Hayne. 2019. "Widespread Shallow Water Ice on Mars at High Latitudes and Midlatitudes." *Geophysical Research Letters* 46: 14290–8.
Planetary Society. 2022. "Space-Warping Planets, the Microlensing Method." https://www.planetary.org/articles/space-warping-planets-the-microlensing-method.
Schmidt, F., F. Andrieu, F. Costard, M. Kocifaj, and A. Meresescu. 2017. "Formation of Recurring Slope Lineae on Mars by Rarefied Gas-Triggered Granular Flows." *Nature Geoscience* 10 (4): 270–3.
Schulze-Makuch, D., R. Heller, and E. Guinan. 2020. "In Search for a Planet Better than Earth: Top Contenders for a Superhabitable World." *Astrobiology* 20 (12): 1394–404.
Seager, S. 2014. "The Future of Spectroscopic Life Detection on Exoplanets." *PNAS* 111 (35): 12634–40.
Seager, S., W. Bains, and J. J. Petkowski. 2016. "Toward a List of Molecules as Potential Biosignature Gases for the Search for Life on Exoplanets and Applications to Terrestrial Biochemistry." *Astrobiology* 16 (6): 465–85.
Shiltsev, V. 2014. "The 1761 Discovery of Venus' Atmosphere: Lomonosov and Others." Journal of Astronomical History & Heritage 17 (1): 85–112.
Smith, P. H., L. K. Tamppari, R. E. Arvidson, D. Bass, D. Blaney, W. V. Boynton, A. Carswell, D. C. Catling, B. C. Clark, T. Duck, E. DeJong, D. Fisher, W. Goetz, H. P. Gunnlaugsson, M. H. Hecht, V. Hipkin, J. Hoffman, S. F. Hviid, H. U. Keller, S. P. Kounaves, C. F. Lange, M.

T. Lemmon, M. B. Madsen, W. J. Markiewicz, J. Marshall, C. P. McKay, M. T. Mellon, D. W. Ming, R. V. Morris, W. T. Pike, N. Renno, U. Staufer, C. Stoker, P. Taylor, J. A. Whiteway, and A. P. Zent. 2009. "H_2O at the Phoenix Landing Site." *Science* 325 (5936): 58–61.

Stefano, R., J. Berndtsson, R. Urquhart, R. Soria, V. Kashyap, T. Carmichael, and N. Imara. 2021. "A Possible Planet Candidate in an External Galaxy Detected Through X-Ray Transit." *Nature Astronomy* 5 (12): 1297.

Struve, O. 1952. "Proposal for a Project of High-Precision Stellar Radial Velocity Work." *Observatory* 72 (870): 199–200.

Turbet, M., J. Leconte, F. Selsis, E. Bolmont, F. Forget, I. Ribas, S. N. Raymond, and G. Anglada-Escudé. 2016. "The Habitability of Proxima Centauri b. II. Possible Climates and Observability." *Astronomy & Astro-Physics* 596: A112.

Walker, S., W. Bains, L. Cronin, S. DasSarma, S. Danielache, S. Domagal-Goldman, B. Kacar, N. Y. Kiang, A. Lenardic, C. T. Reinhard, W. Moore, E. W. Schwieterman, E. L. Shkolnik, and H. B. Smith. 2018. "Exoplanet Biosignatures: Future Directions." *Astrobiology* 18 (6): 779.

Witze, A. 2022. "How NASA's Mars Rover Hit a Geological Jackpot." *Nature* 603 (7899): 18.

Wunderlich, F., M. Scheucher, M. Godolt et al. 2020. "Distinguishing Between Wet and Dry Atmospheres of TRAPPIST-1 e and f." *The Astrophysical Journal* 901: 126.

Zhao, Y., H. Norouzi, M. Azarderakhsh, and A. AghaKouchak. 2021. "Global Patterns of Hottest, Coldest, and Extreme Diurnal Variability on Earth." *Bulletin of the American Meteorological Society* 102 (9): E1672–81.

Appendix 1: Commonly Used Physical and Astronomical Quantities

In astronomy we often find ourselves moving between CGS units (e.g., for astronomy and optics) and MKS units (e.g., for electromagnetics and engineering). Here we list often used constants in both systems.

Speed of light: $c = 2.9979 \times 10^{10}$ cm sec^{-1} = 2.9979×10^{8} m sec^{-1}

Boltzmann constant: $k = 1.3805 \times 10^{-16}$ erg K^{-1} $= 1.3805 \times 10^{-23}$ joule sec^{-1}

Planck constant: $h = 6.6256 \times 10^{-27}$ erg sec = 6.6256×10^{-34} m^{2}kg sec^{-1}

Electron charge: $e = 4.8030 \times 10^{-10}$statC (or esu) = 4.8030×10^{-10} $g^{\frac{1}{2}}cm^{\frac{3}{2}}sec^{-1}$

$= 1.6022 \times 10^{-19}$C

Electron mass: $m_e = 9.1091 \times 10^{-28}$g = 9.1091×10^{-31}kg

Proton mass: $m_p = 1.6725 \times 10^{-24}$g = 1.6725×10^{-27}kg

Atomic mass unit: amu (or u) $= 1.6605 \times 10^{-24}$g $= 1.6605 \times 10^{-27}$kg

Bohr radius: $a_0 = 5.2918 \times 10^{-9}$cm $= 5.2918 \times 10^{-14}$km

Stefan–Boltzmann constant: $\sigma = 5.670 \times 10^{-5}$erg cm^{-2}sec^{-1}K^{-4} $= 5.670 \times 10^{-8}$W m^{-2}K^{-4}

Gravitational constant: $G = 6.6738 \times 10^{-8}$g^{-1}cm^{3}sec^{-2} $= 6.6738 \times 10^{-11}$kg^{-1}m^{3}sec^{-2}

Gravitational acceleration: $g = 9.8067$ m s^{-2} $\Rightarrow$ 32.2 ft s^{-2}

Mass of Sun: $M_\odot = 1.9891 \times 10^{33}$g $= 1.9891 \times 10^{30}$kg

Radius of Sun: $R_\odot = 6.955 \times 10^{10}$cm $= 6.955 \times 10^{8}$m

Luminosity of Sun: $L_\odot = 3.846 \times 10^{33}$ergs sec^{-1} $= 3.846 \times 10^{26}$W

Effective temperature of Sun: $T_e = 5778$ K

Other Quantities:

1 AU $= 1.496 \times 10^{13}$cm $= 1.496 \times 10^{8}$km

1 light year (ly) $= 9.461 \times 10^{17}$cm $= 9.461 \times 10^{14}$km $\Rightarrow 5.879 \times 10^{12}$miles

1 pc $= 3.0857 \times 10^{18}$cm $= 3.0857 \times 10^{13}$km

1 yr $= 3.1557 \times 10^{7}$ sec

1 eV $= 1.6022 \times 10^{-19}$joules

Appendix 2: Exoplanets within 15 Light Years[a]

(All Planetary Detections Made Using Doppler Spectroscopy)

Host Star	Distance	Spectral	Mass	Planet	Planet Mass	Planet Radius	Orbital Radius	Orbital Period	Discovery
	(ly)	Type	($M_{\odot}$)	Label	($M_{\oplus}$)	($R_{\oplus}$)	(AU)	(Days)	(Year)
Proxima Centauri	4.247	M5.5V	0.123	d	> 0.26	0.81	0.029	5.122	2020
				b	> 1.2	1.3	0.049	11.2	2016
				c	7	—	1.489	1,928	2020
Wolf 359	7.856	M6.5V	0.09	c	> 3.8	—	0.018	2.687	2019
				b	> 43.9	—	1.845	2,940	2019
Lalande 21185	8.304	M2V	0.46	b	> 2.7	—	0.079	12.9	2017
				c	> 24.7	—	3.1	3,190	2021
Epsilon Eridani[v]	10.489	K2V	0.781	b	248	—	3.48	2,692	2000
Lacaille 9352	10.724	M0.5V	0.489	b	> 4.2	—	0.068	9.26	2019
				c	> 7.6	—	0.12	21.8	2019
Ross 128	11.007	M4V	0.168	b	> 1.4	—	0.05	9.87	2017
Struve 2398 B[b]	11.491	K5	0.248	b	> 15.7	—	0.261	91.3	2019
				c	> 13.1	—	0.428	192	2019
Groombridge 34 A[b]	11.619	M1.4V	0.38	b	> 3.03	—	0.072	11.4	2014
				c	> 36	—	5.4	7,800	2018
Epsilon Indi A[v]	11.867	K5V	0.762	b	1,030	—	11.55	16,500	2018
Tau Ceti[v]	11.912	G8V	0.78	g	> 1.75	—	0.133	20	2017
				h	> 1.8	—	0.243	49.5	2017
				e	> 3.9	—	0.538	163	2017
				f	> 3.9	—	1.33	640	2017
Gliese 1061	11.984	M5.5V	0.113	b	> 1.4	—	0.021	3.2	2019
				c	> 1.7	—	0.035	6.69	2019
				d	> 1.6	—	0.052	12.4	2019
YZ Ceti	12.122	M4.0V	0.29	b	> 0.75	—	0.0156	1.97	2017
				c	> 1.2	—	0.0209	3.06	2017
				d	> 1.1	—	0.0276	4.66	2017

(Continued)

Host Star	Distance	Spectral	Mass	Planet	Planet Mass	Planet Radius	Orbital Radius	Orbital Period	Discovery
	(ly)	Type	$(M_\oplus)$	Label	$(M_\oplus)$	$(R_\oplus)$	(AU)	(Days)	(Year)
Luyten's Star	12.348	M3.5V	0.29	c	> 1.2	—	0.0365	4.72	2017
				b	> 2.2	—	0.09	`8.6	2017
				d	> 10.8	—	0.712	414	2019
				e	> 9.3	—	0.849	542	2019
Teegarden's Star	12.497	M7.0V	0.08	b	> 1.1	—	0.0252	4.91	2019
				c	> 1.2	—	0.0443	11.4	2019
Wolf 1061	14.05	M3.5V	0.25	b	> 1.9	—	0.376	4.89	2015
				c	> 3.6	—	0.089	17.9	2015
				d	> 6.5	—	0.421	184	2015
Gliese 83.1	14.578	M4.5V	0.14	b	> 30.9	—	0.403	242	2019
				c	> 71.6	—	0.87	768	2019
Gliese 687	14.839	M3.5V	0.41	b	> 17.2	—	0.163	38.1	2014
				c	> 16.0	—	1.165	728	2019
Gliese 674	14.849	M3V	0.35	b	> 11.2	—	0.039	4.69	2007
Gliese 876	15.238	M4V	0.33	d	6.8	—	0.0208	1.94	2005
				c	230	—	0.133	30.2	2000
				b	720	—	0.213	61	1998
				e	15	—	0.342	125	2010

[a] "NASA Exoplanet Archive—Confirmed Planetary Systems". NASA Exoplanet Science Institute. California Institute of Technology. Retrieved 2022-02-19.
[v] Visible to naked eye;
[b] binary system.

Appendix 3: Timeline of Cultural/ Technical Evolution

When	Who	What
3800 BC	Babylonian astronomers/ priests	Accurate Astronomical Calendar
1500	Egyptian astronomers/priests	Star charts
400	Democritus	Atomic theory
350	Plato, Aristotle	Geocentric Model (concept)
230	Aristarchus	Heliocentric Model (concept)
100	Unknown Ancient Greek	Antikythera mechanism; early analog computer for astronomy
200 AD	Ptolemy	Geocentric Model (mathematical)
275	Palmyrene Invaders	Destruction: Library of Alexandria
420	St. Augustine (priest)	Curiosity is a vice to be avoided
1250	St. Thomas of Aquinas (priest)	Scientific inquiry is acceptable, but will lead to metaphysical, not empirical, conclusions.
1440	Johannes Gutenberg	Printing Press
1542	Nicolaus Copernicus	Heliocentric Model (mathematical: circular orbits)
1600	Giordano Bruno (friar)	Heliocentric Model (burned at the stake by Inquisition in Rome)
1601	Tycho Brahe	Greatest naked eye astronomer
1609	Galileo Galilei	Telescopic Astronomical Observations
1615	Johannes Kepler	*Epitome Astronomiae Copernicanae:* Heliocentric Model (mathematical: elliptical orbits)
1687	Isaac Newton	*Philosophiæ Naturalis Principia Mathematica:* Theory of Gravity; Classical Mechanics; Newton interned at Westminster Abbey in 1727
1799	Simon Laplace	*Mécanique celeste:* No Divine intervention needed to explain natural phenomena
1826	Joseph Niepce	First photograph: invented photography
1859	Charles Darwin	*On the Origin of the Species*: discusses natural selection
1861	James Clerk Maxwell	Equations of electromagnetism
1862	Gustav Kirchhoff	Laws of spectroscopy

(*Continued*)

When	Who	What
1879	David Edward Hughes	First Radio Demonstration; Heinrich Hertz was the first to publish in 1887
1882	Louis Pasteur	Disproof of spontaneous generation
1900	Max Planck	Quantization of light
1905	Albert Einstein	Special Theory of Relativity
1908	Henrietta Leavitt	Cepheid-period–luminosity relation
1915	Albert Einstein	General Theory of Relativity
1920	Arthur Eddington	Nuclear fusion in stars
1922	Alexander Friedmann	Predicted expansion of universe through General Relativity
1927-29	Lemaitre, Hubble, Humason	Observed expansion of the universe
1939	John Vincent Atanasoff	First Electronic Digital Computer
1952	Urey and Miller	Urey-Miller Experiment creating amino acids in the laboratory
1953	Watson and Crick	Discovery of DNA
1959	Cocconi and Morrison	*Searching for Interstellar Communications*: Foundation of SETI
1960	Frank Drake	Project Ozma: First SETI Experiment
1973	Drake and Sagan	Arecibo message: first intentional interstellar transmission
1976	Viking Missions to Mars (NASA)	Two of three life detection experiments negative; results remain inconclusive
1995	Queloz and Mayor	Observation of an exoplanet around a main sequence star (51 Pegasus)
2009	William J. Borucki	Principal Investigator of Kepler Space Telescope; Detection of exoplanets with transit technique

Appendix 4: Periodic Table of the Elements*

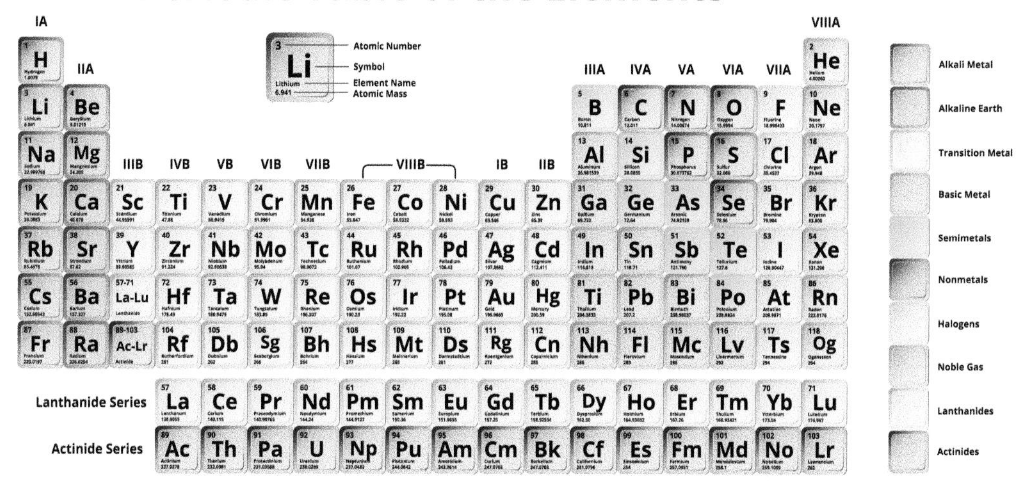

Appendix 5: Required Lifetime of Communicable Civilizations

Since all forms of electromagnetic waves (e.g., radio, infrared, visible, ultraviolet, X-rays, or gamma rays) travel at the speed of light, the average lifetime, L_c, required for civilizations to establish two-way communications is simply,

$$L_c \geq 2r_s \tag{A5.1}$$

where r_s is the distance between communicable civilizations expressed in light years (ly). Expressions for r_s were derived in Chapter 1 for two cases: (1) the neighboring civilization is located in a spherical volume from Earth with a diameter equal to the thickness of the galactic disk, ~1,000 ly and (2) the civilization exists at a further distance within the disk of the Milky Way. For convenience, the associated expressions are provided here.

$$r_S = \left\langle \begin{array}{l} \left(\frac{3}{4\pi}\frac{N_{MW}}{n_*}\right)^{\frac{1}{3}}\left(\frac{1}{N_c}\right)^{\frac{1}{3}} \text{for } r_S \leq 1{,}000\,\text{ly} \\ \left(\frac{1}{\pi h_d}\frac{N_{MW}}{n_*}\right)^{\frac{1}{2}}\left(\frac{1}{N_c}\right)^{\frac{1}{2}} \text{for } r_S > 1{,}000\,\text{ly} \end{array} \right| \tag{A5.2}$$

Let us assume case 1 is correct. Substitution then yields,

$$L_c = 2r_s = 2\left(\left(\frac{3}{4\pi}\right)\left(\frac{N_{MW}}{n_s}\right)\right)^{\frac{1}{3}} N_c^{-\frac{1}{3}} \tag{A5.3}$$

where

L_c = minimum required lifetime of communicable civilization (yr)
H = thickness of the galactic disk (ly) = 1,000 ly
N_{MW} = number of host stars in Milky Way = 4×10^{11} stars
n_s = number density of Sun-like stars $= f_g n_*$ ly^{-3}

n_* = number density of stars per cubic ly = 0.004 ly^{-3}
f_g = fraction of stars that are good stars ≈0.12
N_c = number of communicable civilizations in Milky Way

The Drake Equation can provide us with an estimate for N_c.

$$N_c = \alpha L_c \tag{A5.4}$$

where

$$\alpha = R_* f_g f_p n_e f_l f_i f_c$$

R_* = the star formation rate per year
f_g = fraction of stars that are "good" stars
f_p = fraction of stars that form planets
n_e = number of planets per stellar system that are "Earth-like"
f_l = fraction of planets where life originates
f_i = fraction of planets with life that evolves an intelligent species
f_c = fraction of intelligent species that have the desire and ability to engage in interstellar communications
L_c = minimum required lifetime of interstellar communicative phase in years.
Combining Eq. (A5.4) with (A5.3) and simplifying yields,

$$L_c = \left(2\left(\left(\frac{3}{4\pi}\right)\left(\frac{N_{MW}}{n_s}\right) \right)^{\frac{1}{3}} \alpha^{-\frac{1}{3}} \right)^{\frac{3}{4}}. \tag{A5.5}$$

From the substitution of known astronomical quantities we find,

$$L_c \approx 6{,}264\, \alpha^{-\frac{1}{4}}. \tag{A5.6}$$

The parameter α can be expressed as,

$$\alpha = n_e^{MW} f_l f_i f_c. \tag{A5.7}$$

where
$n_e^{MW} = R_* f_g f_p n_e =$ total number of Earth-like planets in the Milky Way.
From observations taken during the Kepler mission, the current best estimate for n_e^{MW} is ≤6 billion (Kunimoto and Matthews 2020). Substitution then yields,

$$L_c \geq 22.5\left(f_l f_i f_c\right)^{-\frac{1}{4}} \tag{A5.8}$$

Until life is created in the laboratory and/or discovered beyond Earth, estimates of the remaining unknowns, f_l, f_i, and fc, will remain highly speculative. For the sake of argument, let us be somewhat pessimistic and assume each has a one in a hundred chance of occurring, i.e., each has a fractional value of 0.01. Evaluation of Eq. (A5.8) then provides an estimate of the average required minimum lifetime of two neighboring technical civilizations to make two-way contact via electromagnetic waves (e.g., radio) of >700 years. The exponential nature of Eq. (A5.8) makes the resulting estimate relatively insensitive to order of magnitude changes in our most uncertain parameters (f_l, f_i, and fc). This estimate also makes the somewhat optimistic assumptions that the neighboring civilizations are coeval and have the resources and desire to constantly monitor and respond to incoming transmissions from all nearby planetary systems with a Sun-like star. Referring back to Eq. (A5.1), we find the corresponding distance between the home worlds of neighboring civilization to be > 350 ly. If this is the case, we will likely need to continue our SETI efforts over multiple generations, unless our technologically advanced neighbors have anticipated our ascendance and already initiated an attempt at communication (Bialy and Loeb 2018).

REFERENCES

Bialy, S., and A. Loeb. 2018. "Could Solar Radiation Pressure Explain 'Oumuamua's Peculiar Acceleration?" *The Astrophysical Journal Letters* 868: L1.

Kunimoto, M., and J. Matthews. 2020. "Searching the Entirety of Kepler Data. II. Occurrence Rate Estimates for FGK Stars." *AJ* 159: 248.

Index

D

E

F

G

Q

R

S

T

U

V

W

Z